DESIGNING WITH LIGHT

Retail Spaces

Lighting solutions for shops,
malls and markets

DESIGNING WITH LIGHT

Retail Spaces
Lighting solutions for shops, malls and markets

JANET TURNER

Series Editor: Conway Lloyd Morgan

RotoVision

A RotoVision Book
Published and distributed by RotoVision SA
Rue du Bugnon 7
CH-1299 Crans-Près-Céligny
Switzerland

RotoVision SA, Sales & Production Office
Sheridan House, 112–116a Western Road, Hove
East Sussex BN3 1DD, UK

Tel: +44 (0)1273 72 72 68
Fax: +44 (0)1273 72 72 69
E-mail: Sales@RotoVision.com

Distributed to the trade in the United States by
Watson-Guptill Publications
1515 Broadway
New York, NY 10036

ISBN 2-88046-334-3

Book design by Lovegrove Associates

Production and separations in Singapore by
ProVision Pte. Ltd.

Tel: +65 334 7720
Fax: +65 334 7721

Acknowledgements

The author, editor and publisher wish to thank the following for their help, advice and encouragement in the creation of this book: Will Alsop, Paul Atkinson, Brian Bale, Max von Barnholt, Gert Boven, Maurice Brill, Richard Bryant, Clark Crawford, Priscilla Carluccio, Jonathan Court, Steven Dean, Rasheed Din, Paul Dyson, Terry Farrell, Piers Gough, Paul Hannegraf, John Harvey, Robert Heritage, Eva Jiricna, John Johnson, Theo Kondos, Charlotte Kruk, Tony Lawrence, John Edward Linden, John Miescher, Jean Nouvel, Julian Powell-Tuck, Malcolm Robertson, Claude Silvestrin, Robb Stuart, Cynthia Turner, Bob Venning, Edwin Walvisch, Simon White, Barrie Wilde, Julian Wyckham, Terence Woodgate, as well as the following clients, and design, architecture and lighting companies and partnerships for their help: Arup Associates, Ove Arup and Partners, Aspreys, BDP, Bally Switzerland, Hugo Boss, Concord Lighting, Carlton Hobbs, The Conran Shop, Davie Baron, DIN Associates, Erco Lighting, FRCH Design Worldwide, G.E.C. Hong Kong, Into Lighting, Joseph, Calvin Klein, John Lewis Partnership, Liberty, Lighting Design Partnership, Lightcube Japan, Lumiance, Marks & Spencer, Muji, Nissan UK, Paffendor Inc., Gesellschaft Cologne, Racing Green, Regent Switzerland, Spiers and Major, SLI Erlangen and Brussels, Stanton Williams, Selfridges, Stockman Mannequins, Warner Brothers Movie World.

Very special thanks go also to Keith Lovegrove for his work on the design, and to Natalia Price-Cabrera for her editorial vigilance.

Janet Turner
London, March 1998

The photographs reproduced in this book are copyright. We would like to thank the following photographers for permission to reproduce their work: Richard Bryant/Arcaid, Peter Cook, John Edward Linden, Paul Raesied, Philippe Ruault, Malcom Robertson, Andy Whale.

Section One:

Section Two:

On the illustrations in this book, the symbols below describe the type of directional lighting used, with the beam widths *(facing)* as appropriate. These symbols relate to the main lighting system in the relevant illustration.

Direction of Light

 Downlighting   Uplighting   Sidelighting   Spotlighting    Multi-directional lighting

CONTENTS

Section Three:

Section Four:

Beam Widths

 Narrow beam
 Medium beam
 Broad beam

Whether it's facing on to the city high street, or the crowded mall, the quiet courtyard or the airport departure terminal the shop window makes a good starting point for the understanding of the role of lighting in the retail world. This is not a case of beginning in the middle. Rather it is beginning at the edge. The shop window acts both as a frontier, separating street from retail space, and as a visual entrance, displaying the goods on sale, often allowing a visual link through into the shop itself, as well as making a statement about the kind of shop it fronts.

This statement is not merely about the kind of goods on sale, it is also about the attitude, atmosphere, and even ethos of the shop. So getting the lighting of the shop window correct is an important first step in understanding how lighting can support and enhance a retail activity. Let us

look at a number of different approaches to the challenge of the window. One of the first distinctions might be between presentation and invitation. Is the window to be a treasure chest of goods on sale, or is it to show what is happening within? In both cases the intention is the same: to catch the eye of the passer-by, display what is available, and so gain business.

The enclosed shop window is still the hallmark of the traditional department store: they enable changing themes and ideas, and new lines and collections. Their success depends on the inventive skills of the display designer, which can be enhanced by the careful use of light. If the light fittings are concealed, that is, not visible to the onlooker beyond the window, the shop window becomes a small stage set, in which a dramatic lighting solution can be particularly effective.

A traditional approach to the city shop is the small window protected by an awning, as at Calvin Klein in Paris *(left)*: a more recent approach is a fully glazed window, as at Marks & Spencer's Hong Kong store *(above)*.

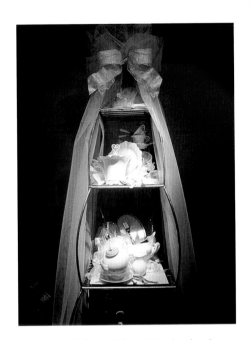

These two windows at Liberty in London show how an enclosed window can be lit dramatically: the light sources, behind a baffle, are not visible, but the lighting effect well planned.

Contemporary approaches to retailing often favour a completely transparent floor-to-ceiling entrance. This can either be left completely clear so that the whole of the shop interior becomes the shop window, or can be partly filled with displays beyond which the rest of the interior is visible.

Ideally, all enclosed shop windows should have at least two alternative lighting circuits, one for daytime, one for night-time. More light within the window is required during daylight to counteract the bright outdoor light from the street, otherwise there will be endless reflections obscuring the view of what is within. Less light is needed after dark. A sophisticated control system can not only solve this problem, but allows for the creation of a number of 'scenes', with changing patterns of light distribution and light colour, creating a real eye-catcher for a special promotion.

A shop specialising in smaller items such as shoes and accessories can, as here, at Bally in Hong Kong, use a low-level display in the front of the window with an overview beyond. The displays in the main area of the shop are lit by recessed ceiling fittings above *(above)*.

At this Shu Uemura shop in a Japanese mall there is no specific lighting for the window, only backlighting for the signage. The strong fluorescent lighting within the shop was chosen to accompany the direct, almost plain packaging for their cosmetic products *(left)*.

The Christmas window is a traditional feature of many Western department stores. This elaborate example from Selfridges in London uses a reproduction Fabergé egg as a centrepiece in a lavish setting *(above)*.

The deliberately dramatic lighting on realistic mannequins highlights the contemporary approach of this fashion window. The absence of background detail, and the use of narrow beams of light falling at an angle on the figures complete the effect *(left)*.

Often the question of the window type will be decided when the shop is built, so the lighting designer will have to work with the space available. The next task is to develop a suitable metaphor for the goods on offer and the general marketing approach or branding of the shop itself. Gold colouring is redolent of the luxury of the traditional department store; it would not be appropriate for a sports shop or music shop, for example.

Is the shop to be seen as being formal, classical, jazzy, contemporary, or cool? How the shop window is lit gives the potential customer an immediate impression of what is to be found within, not merely in terms of goods, but also in terms of the image of the shop.

Traditional images of luxury can, as here, be given a contemporary feel by the use of coloured lighting and contemporary mannequins *(left)*.

The shallow shop window is always difficult to light effectively. The trick is to use lighting at an angle, rather than directly over the mannequins or display items. Crosslighting on the models, and direct and cross downlighting on the boxes behind creates an interesting effect *(above)*. Where overhead downlighting alone is used, the objects are not as well lit *(left)*.

The positioning, branding and intended market of the retail outlet is nowadays the subject of intense research and development. The lighting designer – together with the architects and interior designers – needs to be aware of all the nuances of the market role of the particular shop they are working on in order to arrive at the best solution. This does not merely apply to the shop window, but runs through the whole interior design, through the graphics down to labels, price tags, and bills, and of course the lighting designer plays a key role in implementing this. Thus the lighting designer wishing to work in retail needs to have as broad a knowledge as possible of all the design and marketing aspects of the retail environment, and so I make no apology for bringing into this discussion, and into the pages that follow, considerations of the wider design aspects of any retail project.

In the following chapters you will find firstly an introduction to the basics of light and lighting technology. This is a field in which innovation is the norm. New lamps, new fittings, and new control systems are being introduced to the professional market with ever-increasing rapidity. Understanding how to make a selection among them for a specific lighting task is a key question for the architect, interior designer and lighting designer.

After the technical introduction there are chapters on different retail situations, grouped around existing examples, and with case studies, so that you can see how architects, interior designers and lighting designers around the world have dealt with the challenges of small, large and mega spaces intended for retailing.

This sequence of images *(above)* shows how changes in the direction and distribution of light can create interest and drama in a window display.

Light is visible energy: the primal form of energy, and the measure of all energy if Einstein's famous formula is to be believed, which states that energy is equivalent to mass multiplied by the square of the speed of light. Light is a form of electromagnetic radiation, which is visible to the human eye in a narrow band between red and blue, which lies between 400 and 800 nanometres in wavelength.

Shorter wavelengths include infrared light, longer ones ultraviolet light. Light is what enables us to see. The human eye reacts to light in two ways: within the eye, nerve sensors, called rods, detect the intensity of light, and other sensors, called cones, analyse the colour of perceived objects into a mixture of red, yellow and blue tints. This information is transmitted to the brain, where the image of what we perceive is recreated 'in the mind's eye'.

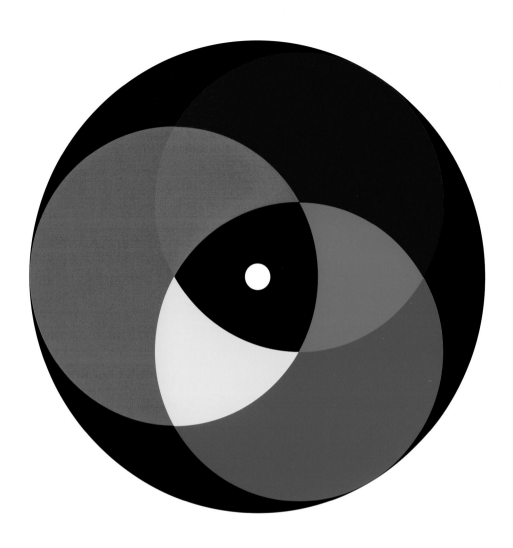

To prove to yourself that colour images are created in the mind, focus on the colour wheel *(left)* for a minute or so, then look at the black dot below. It will appear to be surrounded by colours. But look at the black dot alone and you will see it is surrounded by white. This is part of a phenomenon called persistence of vision, which allows images to be retained in the mind even if the visual information is changing. It allows us to watch and 'read' images on film and television, for example.

The definitions of colour are normally divided into three categories: hue – the overall colour (red, green, orange, for example); value – the overall lightness or darkness of the colour; and chroma – the strength of the colour, its degree of purity or dilution. Thus a strong dark red would be described as having red as a hue and a high chroma and value, a pale washed-out blue as having blue as a hue, and a low chroma and value. Various systems exist for classifying colours along these lines, of which the Munsell system is probably the best known. This gives colours a three figure definition according to the levels of hue, chroma and value.

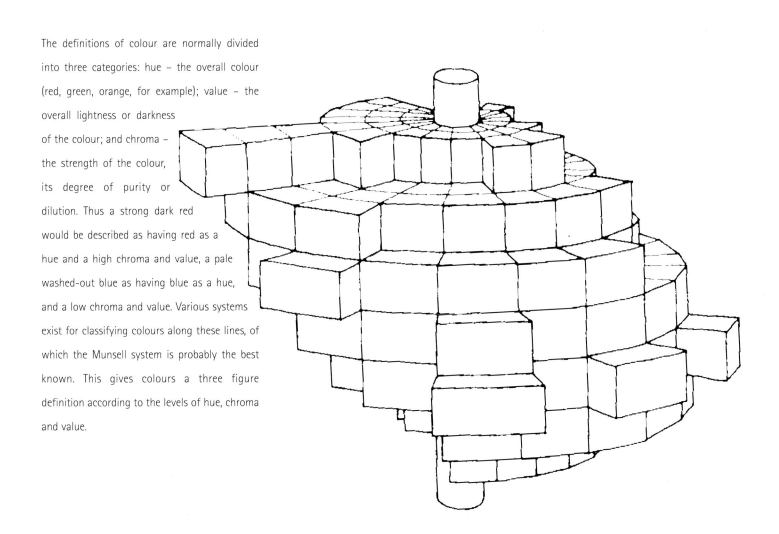

Munsell's colour shape *(above)* is a three-dimensional colour model with a central axis representing chroma, and projections from it representing hue (according to position around the axis, and value according to the length of the projection).

As to the number of visible colours, scientists suggest we can differentiate between some 40 million. In fact, since a wavelength is a ratio, not a fixed quantity, the number of colours in the visible spectrum is theoretically infinite, depending on the number of decimal places the actual wavelength is calculated to. And the number of colours an individual can distinguish depends on the receptivity of the individual's own rods and cones, and is further conditioned by our subjective terminology for colours. One person might describe a particular colour as a greenish blue, another the same colour as a very blueish green. Add to this the fact that many

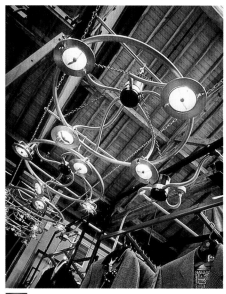

Look at these four interiors: one luxurious and traditional though using modern fittings *(facing, right)*, one functional to the point of minimalism *(below, far left)*, one making use of strong colour to set the mood *(above)*, the last using highly-decorative fittings to create an individual style *(left)*.

colours have, it is claimed, associated psychological effects, which are in turn socially conditioned, and you will realise one of the complexities of working with light: that an objective physical phenomenon that can be measured with great accuracy, is in fact subject to a range of individual, social, cultural and psychological perceptions that may vary greatly from place to place and time to time.

There is an important lesson for designers in this, that the role of lighting is not simply practical, but is affected by a range of other factors to do with the intended users' expectations and perceptions. A successful lighting design is therefore one which not only meets the needs of the brief, but goes beyond it into defining a space in which the users' wider needs are also met.

At the Joan and David fashion shop in New Bond St, London, too much daylight was a problem. The client wanted a warm, strong interior light, so the designer, Eva Jiricna, used slatted stained maple bars to reduce the light from the windows to a gentle flow, supplemented by downlighters over the displays and additional ceiling lighting – a careful balance between natural and artificial light.

Sources of Light

The main source of natural light is sunlight during the day, which is delivered at night in the form of reflected light from the moon, supplemented by starlight. The effect of daylight varies with the weather, the time of day – reflecting the sun's position in the sky – and with the geographical position of the observer, being most intense at the equator and varying in strength and colour balance according to latitude.

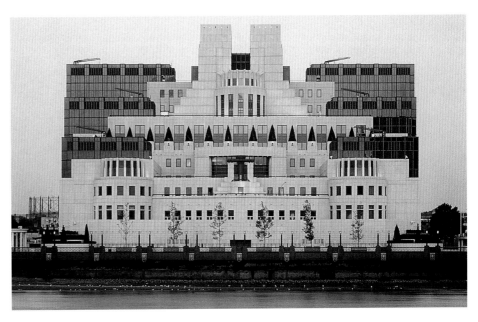

Changes in daylight even in the same country can produce different lighting effects, from the overcast sky of London *(above)* to the bright sky near Bristol *(left)*.

What may be termed 'the daylight constant' is an important factor in individual perception. We each have a kind of baseline perception of what colours and appearances should be, which we recall, subconsciously, under different or impaired lighting conditions. This is important for the lighting designer, since any degree of variance from this norm will have consequences for the appreciation of any lit space for its eventual users. Sometimes this is desirable – no one wants the interior of a nightclub to be lit like a high street at midday, for example – but sometimes it is not – who wants to work in a hospital theatre lit like a nightclub? So awareness of the daylight constant is very necessary for the designer.

The importance of adequate lighting in daylight can be seen from these day- and night-time views of a Nissan car dealership in central London *(right, above and below)*. The interior is lit by ambient light from ceiling downlighters and wallwashing fixtures on separate circuits. There are also two further circuits for accent lighting the cars on display, appropriately to their colours (metal halide lamps for colours from the cool end of the spectrum such as blue and green, white SON for warmer colours such as yellow and red.)

Colour Rendition

When we say that an object is red, what is actually happening is the white light falling on it is absorbed except in the red part of the spectrum, which is reflected. The wavelength of light absorbed, and the intensity of energy reflected in fact depend, in the final analysis, on the chemical composition of the lit surface. We all know for example that a shiny surface, in whatever colour, reflects more light than a matt surface in the same colour.

If there is a change in the colour of the light source, then the reflected colours also change. A common example of this is the sodium lighting used on streets at night, in which the yellow light, while making shapes clearly discernible, flattens their colour range down, so that a blue object will appear black, or a white object yellow. Here our colour memory comes into play, allowing us to evaluate what we see as if it were better lit.

This phenomenon of colour rendition is extremely important for lighting designers. Different surfaces have their own colour values (the range of the colour spectrum they will absorb or reflect). They also have reflectance values (the amount of light – of whatever colour – they reflect). A wall painted in green gloss paint has a known colour value and high reflectance. The same wall covered in a green matt paint may have the same colour value with a much lower reflectance as the surface absorbs more light energy. The important variables for the designer are therefore light colour and reflectance.

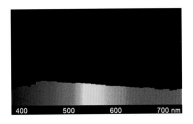

The spectral composition of natural daylight.

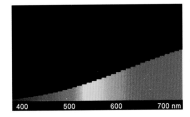

Typical spectral composition of an incandescent lamp showing its bias towards the red end of the spectrum.

27

Colour of Light

Light sources do not emit an even white light, as anyone who has looked at different artificial light sources will know. The best known measure of the colour of light is colour temperature. This takes as a model an ideal surface which changes temperature when heated, and according to the light emitted the temperature is calculated. So a cloudless summer day would have a colour temperature of 10,000 degrees kelvin, a dull afternoon, about 4,000 degrees and a cloudy sky around 6,000 degrees. Colour temperature is therefore a measure of intensity, and the formal analysis of light colour is achieved by analysing the spectral composition of the light from a source by plotting the wavelength of light on a graph. This relationship between colour rendition and temperature can be seen in the fact that low temperature sources tend to be weighted towards the lower, red end of the spectrum, while high temperature sources are towards the blue end. The logic of this is confirmed by our experience; afternoon sunlight seems to have a red tinge, while bright light at noon has a blueish cast.

The lighting designer does not need to know the mathematics or exact methods for calculating these values since they are available in technical manuals from suppliers and lighting associations, but a general understanding of the principles involved in measuring light is a valuable aid in developing a successful design.

Artificial Light

Artificial light, from the candle flame to the low-voltage tungsten dichroic, the triphosphor fluorescent or the simple incandescent globe lamp used in many homes, is produced by using energy – normally electrical – to create light. On the next four pages you will find an analysis of how contemporary lamps use electricity to create light, together with the basic classification of lamp types. The basic principle involves passing electricity through a space or a metal coil within an inert gas or vacuum, which translates that energy into light, and some of the energy into heat. To operate, a lamp needs to be placed in a fitting; this can be a simple lamp-holder or a complex device containing lenses and reflectors to guide and define the output. (Some contemporary lamps also have inbuilt reflectors and diffusers.) A 'light source', or light, is therefore the combination of a lamp (or lamps) and a fitting (and any necessary control gear, such as a transformer).

In deciding on the appropriate light source to use for part or all of a lighting brief, the designer needs to have certain basic considerations in mind. Some of these relate to the practicality of the task, such as output, efficiency and cost, some to the aesthetic appearance of the lit space, such as light distribution, intensity and diffusion. These different elements must be integrated into a final solution which will not depend entirely on lamp types, important as these are, but more often than not on devising a system that uses different lamps and fittings to achieve an overall result.

A downlighter fitted with a glass diffuser ring *(above)* and a low-voltage spotlight with dichroic lamp *(below)*.

29

Lamp Types

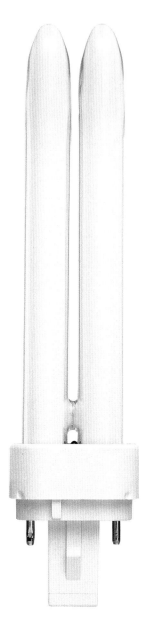

For practical purposes there are three main categories of lamps. These are fluorescent, including compact fluorescent, incandescent, and discharge. The classification depends on the electrical system used to create light, either by passing a current through a wire filament, or through an envelope filled with a reactive gas.

Fluorescent Lamps

Fluorescent lamps operate by passing electrical current through a gas or vapour contained in glass bulb, which is in turn coated with phosphor on the inside. The electrical discharge created within the vapour (often mercury vapour) is transformed into light energy by passage through the phosphor coating. Such lamps are efficient, in that a good proportion of the discharge energy is turned into light, and modern developments which use a triple coating of phosphor have allowed these lamps to give much better colour rendering. Fluorescent lamps also have a long life.

The new range of compact fluorescent lamps offers the performance and life of conventional fluorescents, with the added advantage of a

small lamp size. They are increasingly popular in many design applications, and a wide range of fittings is now available for them. Miniature fluorescents allow excellent lighting from a lamp little bigger than a pencil.

Standard fluorescent lamps require a ballast and a starter gear; today electronic high frequency ballasts should be used for preference, since these both avoid problems of flicker on start-up, and allow the lamps to be dimmed.

Incandescent Lamps

These lamps generate light by the passage of an electric current through a wire coil mounted in a vacuum. These were the earliest type of lamps to be developed, and are still widely available, particularly for domestic use. They are, however, relatively inefficient, much of the energy used being radiated as heat rather than light, and the lamp life is relatively short. The first major development in incandescent lamps was the tungsten GLS lamp, which has a warm colour, followed by the PAR lamp, with its integral

reflector, which allows better directional control. PAR lamps also have a longer lamp life.

Tungsten halogen lamps have a light quality that is closer to daylight than standard or tungsten incandescents, as well as longer lamp life. New developments in mains-voltage tungsten halogen allow these lamps to be retrofitted into existing mains systems.

Lower voltage tungsten halogen lamps offer excellent colour rendition, small lamp sizes, long life and low operating costs. However, these lamps require transformers which are either integral to the fitting or need to be mounted near the lamp.

Two compact fluorescent low voltage lamps *(facing)*. A PAR 20 tungsten mains-voltage lamp *(above)*. A standard mains voltage incandescent lamp with a silvered crown *(right)*.

31

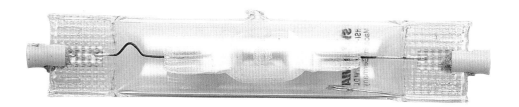

Discharge Lamps

These lamps are considerably efficient, whether in low- or high-pressure models, but require considerable control gear, and cannot always be used in alternative burning positions. High intensity discharge lamps are based on mercury or sodium vapour as the discharge medium; sodium lamps tend to have an orange white light colour and mercury a blueish colour.

Low-pressure sodium lamps emit a flat yellowish light with poor to no colour rendering: they are not appropriate for indoor use.

A double-ended mains voltage lamp *(above)*. A tungsten halogen axial filament low-voltage capsule lamp *(middle)*. A metal halide lamp *(below)*.

New lamps are always being developed: miniature fluorescents, mains-voltage halogen lamps, CDMT lamps, ES 50 lamps and

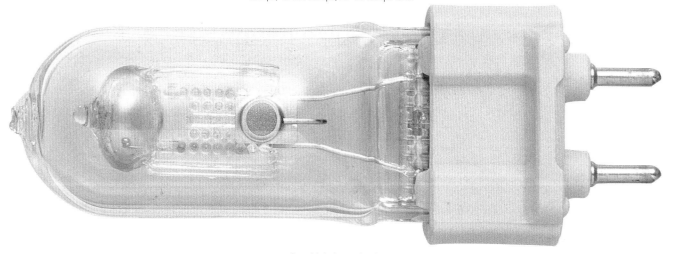

mercury-free high intensity lamps are also now available, the last an important development for ecological reasons.

Summary Guide to Lamp Types

	Type	Wattage	Life in hours	Colour temperature
Mains-voltage (240 volts) tungsten	Standard GLS General purpose incandescent with good colour reproduction.	25–200 watt	1,000	2,700
	Reflector lamps Mirror reflective coating on the inside of the lamp creates directional light in an uniform beam at angles between 25 (narrow) and 80 (broad) degrees.	25–150 watt	1,000	2,700
	PAR 38 Controlled light dispersal (12–30 degree beam angle) with high mechanical strength.	60–120 watt	2,000	2,700
Mains-voltage (240 volts) halogen	QT Capsule Quartz halogen lamps for high output (luminous flux up to 5,000).	75–300 watt	1,500–4,000	2,900
	Halogen DLX Much whiter light than standard incandescent with double the lamp life.	75–100 watt	2,000	2,900
	Halogen PAR 20	50 watt	2,000	2,900
	Halogen PAR 30	75–100 watt	2,500	2,900
	Halogen PAR 38 Can replace standard PAR or reflector lamps with higher output halogen.	75–150 watt	2,500	2,900
	QTY-DE Luminous flux up to 9,500 from a double-ended tube.	150–500 watt	2,000	2,800–2,950
	ES 50 Small footprint but excellent output from integral reflector.	50 watt	2,500	2,900
Mains-voltage (240 volts) fluorescent	Tubular fluorescent Colour temperature depends on lamp colour.	18–70 watt	7,000	2,900–4,300
	Triphosphor fluorescent Triphosphor lamps are recommended for applications where accurate colour values are necessary.	18–70 watt	7,000	2,700–6,000
Mains-voltage (240 volts) compact fluorescent	Lynx s/se Compacts offer the advantages of tubular fluorescent in a smaller format.	5–11 watt	8,000	2,700–4,000
	Lynx d/de, l/le, f These lamps require starter gear.	10–55 watt	8,000	2,700–4,000
	Mini-lynx Contain integral starters, so can be retro-fitted to standard mains lamp-holders.	7–20 watt	10,000	2,700
Low-voltage (12 volts) halogen	QT Compact dimensions with good output and colour rendering.	20–100 watt	2,000–3,000	3,000
	Dichroic Well-defined beam (angles 8–60 degrees), excellent colour rendition, especially useful in food and heat sensitive contexts as 70 per cent of heat generated is radiated backwards.	20–75 watt	2,000	3,000
	Metal reflector Aluminium reflector lamps with precisely directed beam (angles 6–32 degrees).	20–50 watt	2,000	3,000
High-intensity discharge mains voltage (240 volts)	HIT/HIE metal halide Good output with low colour rendering, require starter gear.	35–150 watt	5,000	3,000–4,000
	HIT PAR 38 Metal halide lamp with PAR integral reflector.	100 watt	7,500	3,200
	HST SON Output equivalent to standard GLS lamp.	50–150 watt	12,000	2,100
	CDM	35–150 watt	6,000	3,000

Fitting Types

A light fitting performs a number of functions. It enables the electrical connection to the lamp itself, it protects the lamp, and it directs or diffuses the light from the lamp. In an indoor situation the most important aspect of a fitting is the way in which it controls the flow of light. While fittings can be broadly divided into downlighters, uplighters, spotlights, ceiling-mounted, suspended and recessed fittings, many sophisticated modern fittings can be used in alternative positions, and combine qualities from other types, just as many lamps are configured in ways that in themselves direct the flow of light. A recessed fitting can also be a spotlight, for example, and indeed a downlighter can be recessed into a floor as easily as into a ceiling. (The main consideration here is whether the lamp itself will only operate in a specific burning position, together with the question of heat diffusion around the fitting, and the provision of a safety glass.)

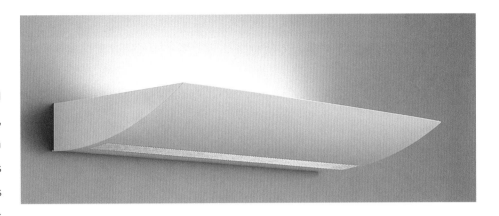

In many cases the fitting can be used with an alternative range of lamps; for example, a standard spotlight fitting will often accommodate lamps with different beam widths where the work of controlling the flow of light is done by the lamp itself. Additional features available for fittings include diffusers, often used over fluorescent lamps, slots on recessed fittings, and plain or tinted glass rings to achieve an alternative diffusion of light from a recessed fitting.

The Torus 100 *(facing, above)* has a low-voltage QT capsule lamp, for which the transformer is housed in the fitting. The reflector is also part of the fitting. Two spotlights that do the same task in different ways. The Lytespot *(facing, below)* has a mains-voltage ES 50 50-watt dichroic lamp with an integral reflector.

A wall-mounted uplighter *(top)* and two downlighters, one with a metal halide lamp *(above, left)* and the other with a dichroic lamp *(above, right)*: note the different apertures required for the different sizes of fitting.

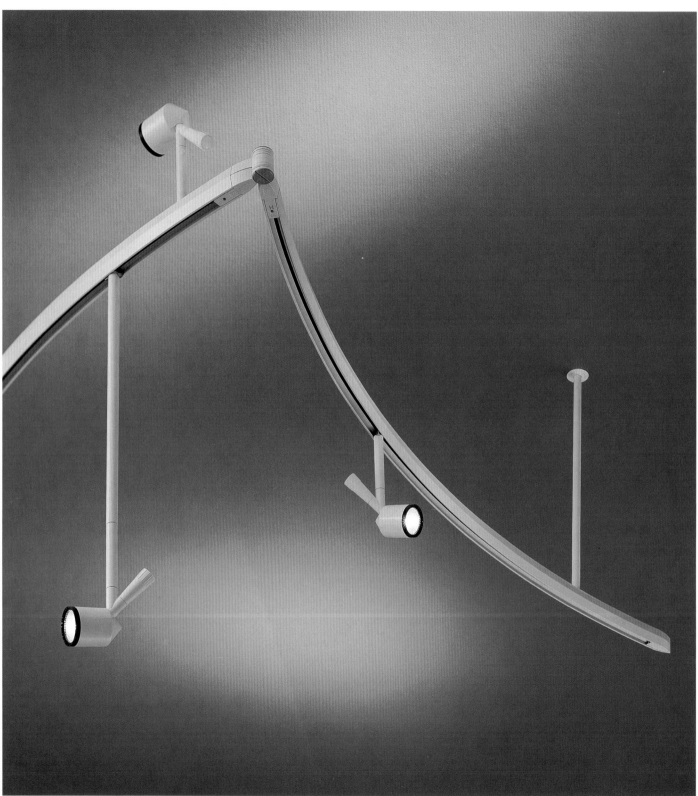

Output, Efficiency and Cost

The output of a lamp is measured in lumens (in the UK and Europe) or candela (in the USA). This is a general measure calculating the light output, and typical values for different lamp types can be found in the chart on page 33. Where a very precise calculation of output is needed, the manufacturer's own documentation should be consulted.

Efficiency refers to the proportion of energy consumed by a lamp that is emitted as light, and also to the ability of the lamp to maintain consistent output. Modern lamps are generally very efficient in this latter respect, maintaining an even output over a longer percentage of their lamp life than was the case earlier.

Cost can be calculated in two ways. Firstly there is the capital cost of a lighting installation, representing the price of fittings, control gear, transformers, lamps, etc. Then there is the running cost of the installation, represented by the electrical consumption, and the costs of relamping. There are various systems of cost comparison, but a rule of thumb is to divide the capital cost by the likely lifespan of the installation and add this to the running cost to establish a yearly total.

Relamping can either be carried out on a formal basis – all the lamps in an installation being changed after a fixed period – or can be done ad hoc, that is to say lamps being replaced as and when they fail. Both systems have their advantages; the main consideration for the designer, as we shall see, is to ensure that access to lamps that may need frequent changing is not too difficult.

The Infinite low-voltage track system allows for endless variations in light distribution *(facing)*. It can be rotated 360 degrees in the horizontal plane and 280 in the vertical plane.

Light Distribution

Light distribution is a function both of the lamp and the fitting, since in most cases the fitting controls the direction and the power of the light source. The designer needs to understand the distribution pattern of a particular fitting and this is best achieved by looking at a diagram called a polar curve for the fitting in question. This plots the outline and intensity of light from a specific fitting.

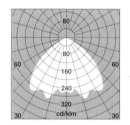

A polar curve diagram represents graphically the luminous output of a given fitting and lamp in different directions. The central column of numerals indicates the intensity in candelas per lumen, the outer columns, the angle of output in degrees. For most fittings a single plane is shown, but for a wallwasher or other highly-directional fittings, two planes, perpendicular to each other are represented.

Intensity and Diffusion

The intensity of light falling on a particular surface, for example a worktop, sales point, or display is translated by looking at the incident light falling on it in lumens. There are guidelines available for the desired level of lighting in different work and retail situations; in some cases these are linked to official regulations, and the designer needs to be aware of these.

Some recommended levels of lighting are as follows: the recommended level of illuminance in lux (one lumen per square metre) for small shops is 500 lux, for supermarkets and hypermarkets it is the same amount, calculated on the vertical faces of displays, and for checkout areas also 500 lux calculated on the horizontal plane of the checkout or sales area. These can be compared with levels of 150 lux, deemed appropriate for areas such as loading bays, 750 lux for drawing offices, and 1,500 lux for hand tailoring, operating theatres and suchlike.

The diffusion of light refers to the overall level of lighting achieved in a scheme by the complete installation. Computer programs are now widely available for calculating these results, and for testing alternative solutions for a specific interior space.

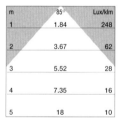

m	85°	Lux/klm
1	1.84	248
2	3.67	62
3	5.52	28
4	7.35	16
5	18	10

A performance cone shows the beam spread from a fitting, the angle being given at the top of the diagram. The left-hand column shows the distance from the fitting in metres, the central column the beam diameter at that distance, and the right-hand column the maximum illuminance (in lux per kilolumen) at that distance.

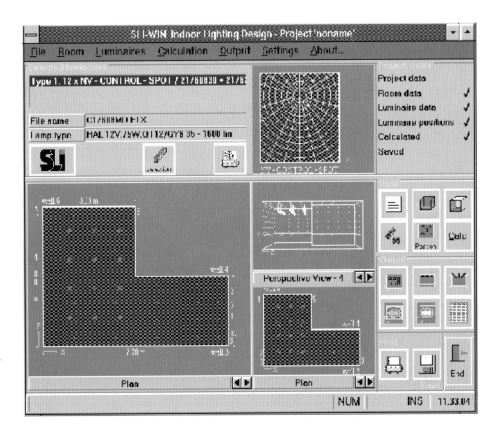

A typical interior program allows lighting levels to be calculated in a rectangular or L-shaped interior space *(above, right)*. The user selects and positions the luminaires in the space from the available database and the program produces a visible output of the lighting distribution *(below, right)*.

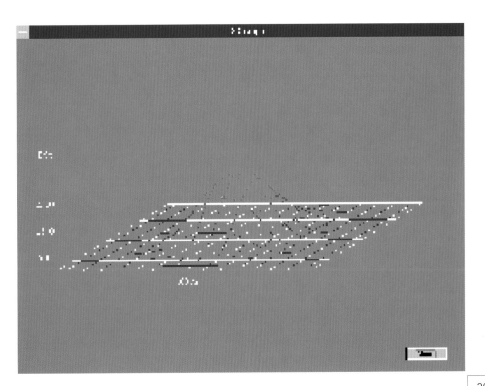

To calculate lighting values exactly, lamp and lighting manufacturers have created programs that will produce an outline of lighting levels in a regular space from exact technical data. These programs allow the architect or designer to check that lighting distribution and levels meet, for example, health and safety requirements. However, such programs do not produce a rendered image, so that the visualisation process is incomplete. High-end graphics programs can create visually appealing images, either from CAD data or from the designer's eye, and technical programs can describe the lighting solutions exactly. What remained to be done was to marry the two up so that the designer could show a client an accurate view of the project.

Two recent programs have squared this particular circle. The first, Lightscape, is a stand-alone program. It offers a series of libraries of surfaces, elements and light fittings that can be assembled to replicate an interior and can then be lit. The complete radiosity program calculates the lighting effect accurately and from all positions. Lightspace recently released a link for 3DStudioMax and the new 3DStudioViz that allows the program to import files from 3DStudio and render them using the Lightscape radiosity engine. Lightscape are also regularly releasing further libraries that contain both architectural elements and specific light fittings to extend and enhance the range of luminaires that can be used under Lightscape.

RadioRay, developed for Autodesk by the UK company LightWorks, is a plug-in for 3DStudioMax and the new 3DStudioViz. This also allows precise light fittings to be inserted into interiors created in 3DStudio, using the industry standard elx file extension. These files, created by lamp and fitting manufacturers, contain complete photogrammetric data allowing the exact performance of a lamp and fitting to be depicted. So in a finalised and rendered design, precise lux values can be read off walls and work surfaces, and the colour and diffusion of light are also visually accurate. Again, since this is a radiosity program, the values are created for all viewpoints.

These two images, of a mall *(facing)* and a supermarket *(above)*, were created entirely using the Lightscape computer program.

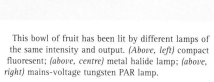

This bowl of fruit has been lit by different lamps of the same intensity and output. *(Above, left)* compact fluoresent; *(above, centre)* metal halide lamp; *(above, right)* mains-voltage tungsten PAR lamp.

Lighting distribution affects the appearance of this floral display. *(Above, left)* general floodlight; *(above, centre)* narrow beam capsule spot; *(above, right)* medium beam dichroic spot.

Light and Visual Effect

The light colour, angle of incidence, and spread of light on an object will have a radical effect on how the object is seen. Something lit directly from overhead looks radically different to the same thing lit from below or at an angle. And when similarly the same object is seen lit by the different light sources (even of comparable outputs), it will appear with very different colour balances and even rendering of detail.

In an uniformly lit space, light models objects according to the lamp colour and output, and according to the colour and reflectance of the lit objects. In a largely spotlit space, the choice of beam width and positioning will also affect the overall visual impression of the space. Thus, a single narrow spotlight will illuminate the centre of a space but leave the edges dim; similarly, a wallwashing light on one wall only, will leave the other sides of the space relatively dark. The lighting designer's task is to achieve overall lighting levels that are visually interesting, but also safe and practical to use.

A linear system comprising a fluorescent behind a louvre, and dichroic spotlights. The design is based on a special fitting created for Birmingham School of Art.

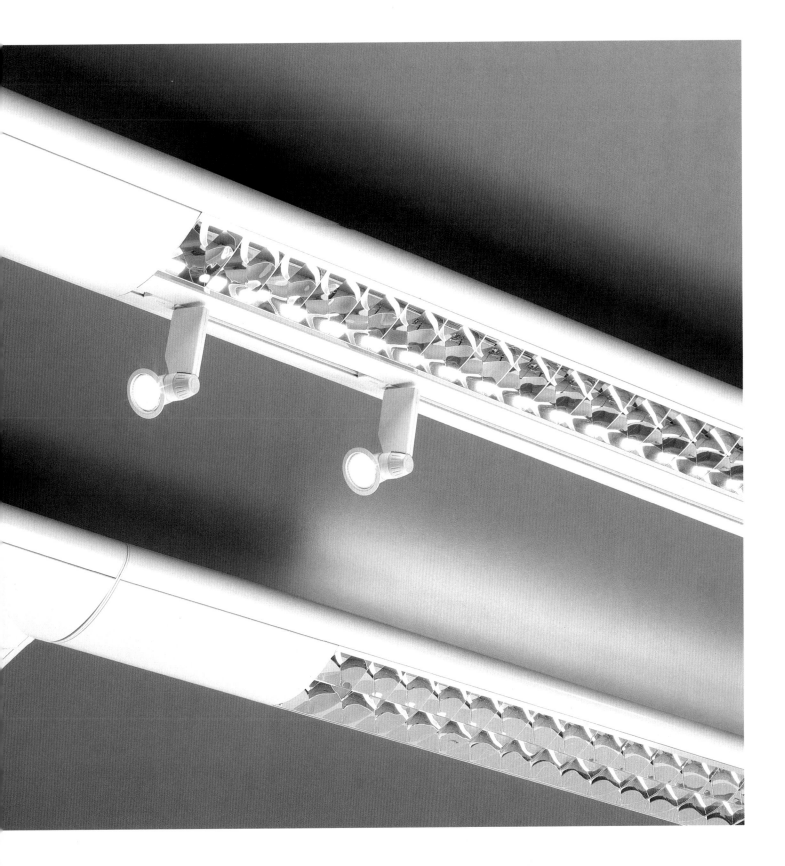

On the kitchen door of a famous contemporary London restaurant there's a notice addressed to the waiters leaving the kitchen. It says very simply 'beyond here you are on stage'. The same might be said, the other way around, of the retail environment; here the customers need to feel that the goods on show are centre stage.

One way of introducing the customer to these objects of desire is through the use of lighting. The designer needs to suggest to the customers that this is the store that they have all been looking for. The lighting design must make the objects of desire even more desirable.

There is considerable research (and even more anecdote) about how people behave while shopping. For example, some of those who study shopping movement patterns have suggested that faced with alternative pathways into a large store interior, most customers will choose to turn right, rather than go left or straight on. (Why this should be so is by no means clear, though right-handedness is one possible factor.) Colour theorists also have views about how combinations of colours reflect different personalities, and that these combinations should be selected as appropriate to the target consumer group.

Entrances and Exits

If the shop window provides a first opportunity to draw the customer's attention, then the entrance is the second step in closing the encounter with the customer: the gateway to the goods on display. The degree of attention the customer is shown or expects is partly a matter of cultural convention (lift attendants, for example, have virtually disappeared from department stores in Europe, but no respectable Japanese store would be complete without its host of welcomers to greet clients). The nature of the products offered also affects how the customer is met: one expects personal service in an exclusive jewellery shop but not in a push-your-own-trolley supermarket! This requirement has an impact, in turn, on the lighting requirements for the entrance space which needs to lead seamlessly into the main display area or areas.

Thus, the lighting designer needs to conceive of the total lighting solution for the shop as a complete system: the emphasis may differ at different points, particularly in a store selling different ranges of goods over a large floor area or several floors, and if, as is so often the case today in larger stores, there are in-store concession sites for particular manufacturer's products. On a large site, these choices can have important consequences for lighting maintenance: one major London department store as a matter of policy only uses eight different lamps, though in a wide range of fittings, so reducing storage space for spare lamps and the risk of incorrect relamping. These practical considerations, as well as questions such as the accessibility of fittings, heat dispersal and so forth, all need to form part of the lighting plan.

Spotlighting the customer: at the Kaufhof store in Frankfurt concentric rings of recessed downlighters – compact fluorescents on the outside, brighter and whiter tungsten halogen on the inner four rings – create a pool of light across the entrance plane.

Getting the ambience right for the intended customers is a major part of the overall marketing strategy of the shop, and it is the lighting designer's task to find the correct solution to meet this requirement. In some cases, the demand will be for a bright and engaging, busy and cheerful approach, or alternatively, for something discreet and cool, or for colour and activity. The solution adopted will not only have to work at the entrance to the shop, but can also be a valuable reinforcement of the strategic marketing approach in intermediate areas such as stairways and escalators. The treatment of these elements, and other architectural features, can help focus customer attention on the central retail experience.

The expression of discretion: the entrance to the Muji Japanese shop in London *(facing, above)*. The lighting is concentrated behind the line of the arch, with the store name etched into the glass and lit from the front, and repeated in a coloured fluorescent sign at the end of the shop. The interior is lit by ambient lighting from compact fluorescents in the relatively high ceiling, supplemented by tungsten halogen accent lighting.

The Gucci entrance (in a mall setting) uses a similar approach, with the shop name behind the door plane *(facing, below)*. The trees in tubs flanking the entrance are an elegant and traditional touch, though a slightly brighter light level would have rendered these more clearly. And an invitation to the customer into a special, almost private place, can be literally conveyed by the use of gates and discreet, recessed downlighters, simulating the daylit space of a real courtyard *(above)*.

Colour plays an important role in shop interior design. It can be used to set the mood, as well as to mark the difference between sections of the shop. The selection of colours needs to be made with the intended merchandise in mind, and the lighting designer should be consulted early in this process as the choice of colours can relate to the suitability of different lamps. For example, metal halide lamps, which have a relatively high colour temperature, are better for lighting dark colours, such as blues and greens, while tungsten halogen, or CMDT lamps, give good colour resolution for softer colours such as reds and yellows. Since both display colours and merchandise colours in a shop are likely to change with time and fashion, it is a good principle to provide for as wide a range of alternatives as the lighting budget will allow. This can be done, for example, by specifying multiple circuit track, rather than single circuit, by choosing spotlights that can be redirected easily, either on track or from fixed positions, and by selecting fittings that can be fitted with different lamps to provide different colour renderings. Many lamp manufacturers are now producing lamps (such as low-voltage compact fluorescents with integral transformers) that can be retrofitted into standard mains fittings.

The use of strong reds and yellows to highlight the entrance to the toy department is accentuated by the use of a strong spotlight on the mannequins beside it *(facing, left)*. An equally strong colour palette is used by DIN Associates in this fashion setting *(above)*. The tones of blue and purple not only mark off the ramp of stairs from the main wooden floor, but also echo the unusual architecture of the roof space.

Floor-mounted uplighters are supplemented by spotlights along the air-conditioning trunking which provide the main lighting, with the recessed area to the left lit by recessed downlighters in the display system. This combined with the natural light through the roof creates a considerable contrast to the track-lit display area at the top of the stairs.

53

The changing levels in the Kenzo shop *(above)* are lit by rows of downlighters in the ceiling and angled wallwashing downlighters around the perimeter. Staircase at the Ted Lapidus store, Caracas, Venezuela *(facing)*.

Staircases offer an opportunity for both architectural and lighting design flair. The change of level and pace offers the customer a new vantage point from the staircase, both visually and mentally. A series of steps can also draw the customer to explore the store, as at the Kenzo shop in London. Spotlights are used to highlight the steel and glass staircase at Bergdorf Goodman in New York, designed by Eva Jiricna, and central downlighters create a white cage at the Ted Lapidus store in Caracas, designed in the late 1960s – good classical design does not date.

At Bergdorf Goodman the staircase is lit from above so that the complex constructional details are clearly visible: a deliberate high-tech statement emphasised by the lighting *(left)*.

Escalators can be both a transport necessity and a visual horror. A necessity, in that they need to be well lit for safety reasons and to clarify a position in the shop. A horror, as the materials from which they are constructed are often highly reflective. But an enterprising designer can turn these features into something positive, either by using coloured or pinpoint lighting, or by using precise reflections from moving parts to animate the setting. In the same way that an entrance can be lit so as to invite the customer in, a functional transport system can be used to keep the customer's interest, maintaining the level of curiosity about the goods yet to be seen.

Conventional striplights are given a hint of difference by the use of pink colour in a department store escalator *(above, left)*. At the London Pavilion the transparent side walls of the escalators, and the highly polished floor create a seemingly endless series of reflections of the pinpoint spots mounted into three banks above each escalator. For the sparkle effect at the Pavilion not to become glaring, the lamp wattage needs to be kept under ten watts per lamp *(above, right)*. In a Hong Kong mall, the chromed facings of the escalator pick up reflections from the main lighting *(below, right)*.

A busy array of shelving invites the customer to browse and select: crowding goods together is fine if the lighting is well planned. Note the positioning of tracks along the aisle centres, with angled floodlights in alternate directions *(above)*.

Shelving and Showing

Entrances get the customer in: the next challenge is displaying the goods. Racks and shelves, hangers and cabinets: the choice will depend on the merchandise on offer, its varieties and sizes, and the buying patterns of the clients. Free access to bulk piles of stock is apropos for a discount warehouse but inappropriate for an *haute couture* fashion shop, for example. One of the ways to resolve this question is to look at the distribution of staff space and customer space. Take the example of a pharmacy: a lot of the products can be on open shelves for the customer to select individually, but a certain number can only be dispensed by the staff. This proportion will dictate the arrangement of the floor area, and so help define the display options. Such an approach can help with other retail solutions, even where there is no safety or legal aspect to consider, as at a pharmacy.

The important question for lighting display fittings is that the goods are what need to be lit, not the fittings. This seems obvious, but it is often overlooked. The way not to do it can be seen in the shelves used for a perfume display in the shop window shown below; quite apart from the fact that the display area is underlit and there is a lot of reflection in the window, the lighting is unbalanced, leaving part of the display in shadow. This, despite the fact that the display is lit from both above and below, the latter not normally an option in a public space. A better approach, seen above in a racking display of towels,

A correct way to light shelving is vertically from above. The distance between the fitting and the shelf edge determines how far down the wash of light will go, while a reflector baffle on the fitting directs the light only on to the shelving *(above, left)*. Using spotlighting, even from below as well as above the shelving, creates an uneven light, and so a poor solution *(below, left)*.

Toplighting above the shelving at the Racing
Green clothes store in London adds to the
effect created by the main lighting to ensure
the shelves are clearly lit *(above)*.

Closed-end shelving inevitably creates interior
shadows that mask what is on display, even where
the display fittings themselves are attractive
(facing, top right), while an open-ended shelving
system allows light to flow through and across the
merchandise *(facing, top left)*. Skilful display systems,
well lit from overhead, can give the appearance of
clothes suspended in space *(facing, below)*.

is to create an even wash of light from above, and sufficiently far in from the display to light it all. Often the right solution depends on combining dedicated light over the shelving with supplementary directional light from the rest of the installation. At Racing Green, this is achieved by wallwashing downlighters over the recessed shelving, and tracked spotlights in the main ceiling.

Closed shelving can also present a problem. The mobile display units in this gift shop shown above are quite elegant and generally well lit from the dark ceiling above, but the closed ends of the shelves create too many interior shadows. One very creative solution is to lose the tops and ends of the shelves altogether, either putting the display on a mobile base, as seen here on the right, or by hanging an open shelf from the ceiling, as shown above.

Display cases are necessary to protect fragile or valuable goods from damage or loss. They can also be used to highlight a particular item or range, a physical expression of its speciality or rarity, as in this softly lit jewellery display seen below. But the same process also denies access to the goods, which can frustrate the customer, so display cases need careful consideration, and demand careful lighting.

More modern solutions, whether in a wall of display boxes *(facing, above)* or with freestanding units *(above)* achieve better results through a range of lamps and fittings. Note in the facing image the use of the back of the box as a reflective surface to redistribute the light from miniature fittings above the items displayed, and how narrow beam spotlights from the ceiling frame each pedestal at the Warner Brothers Movie World, Berlin *(above)*.

For a long time, lighting display cases successfully was a problem, as this glassware example from the mid 1980s shows. Too much light could create endless internal reflections, and also create heating problems within the enclosed space. Too little light left parts of the display in darkness. Modern lamps and fittings, which are both smaller and create less heat, can help overcome this problem, as the curved range of boxes showing antique glassware proves. In particular, fibre optic lamps, where the light source is remote from the point of delivery, can deliver a cool, safe and gentle light.

Colour and contrast can be used to heighten the experience of the customer, even of goods in showcases. At Warner Brothers Movie World, the display cases become the stars through a lighting system that isolates them on the strongly coloured carpet, using just the recessed fluorescent fittings beneath the top lid of the cases over a parabolic low brightness louvre, combined with tungsten halogen narrow beam spots on the walkways between. The result is an almost cinematic atmosphere, in which the goods are well lit.

Welcoming the customer in, and showing what is on offer to best advantage should be the main objectives of a successful lighting design for a retail space. How much light to use, and where to place it, will in part be defined by experience and practice, and in part by the marketing brief for the space. The lighting solution must be a total unity, working in harmony with the whole design statement of the shop.

These two 1980s examples of display cases use mains-voltage tungsten light. This creates a flat wash of light, which may not show goods off to best advantage, even with glass shelving to carry the toplight through *(left)*. Often today a secondary source, usually low voltage, would be included to add sparkle, especially for items such as jewellery *(facing, below)*. The low voltage would also not increase the heat problem.

Some years ago marketing experts used to divide consumers into broad groups with labels such as A, B, AB, C or D. These largely equated income, class, taste and social aspirations between each group. The economic changes of the 1980s, the growth of travel and the emergence of new consumer groups have made a nonsense of these broad brush distinctions.

Now the targetting of products, services and retail outlets is very much more sophisticated. Nowhere is this process to be seen more vividly than in fashion retailing. Competition in a growingly global market puts a premium on positioning, differentiation and distinctiveness. The growth of brands such as Calvin Klein, Paul Smith or Donna Karan beyond not only urban but also national boundaries, the profusion of consumer magazines in fashion and lifestyle, and the relaxation of traditional dress codes in Western culture have all fuelled this development.

So the retail environment is one of the most exciting and challenging for designers, whether dealing with designer labels, *haute couture*, sportswear or casual clothing. In creating appropriate retail environments, the lighting designer is an important member of the team.

The interior of the Calvin Klein shop in Paris, designed by Claude Silvestrin. The adjustable recessed fittings, in modular groupings, deliver light on to the merchandise without ceiling clutter *(facing)*.

The towels and accessories at the Calvin Klein shop in Paris are set around a stoneware bath. Even if the interior design is minimal, a good quantity of light has been included, emitted from fittings housed with appropriate discretion in the ceilings *(left)*.

A gobo (a stencil plate fitted behind a spotlight lens) used to brand the shelving at Caren Pfleger *(left)*.

The level of illuminance within the shop is deliberately greater than the general level outside, ensuring that the window's contents are always clearly visible from the outside *(above)*.

Caren Pfleger Collection, Fashion, Berlin, Germany.

'The style shows the wearer is in touch – with self, with society, with nature.'

One recent trend in clothing is towards informal clothes in soft, natural colours, both for men and women. They relate to a caring, conscious lifestyle: neither formal nor showy, and balanced, not excessive. The style shows the wearer is in touch – with self, with society, with nature. The Caren Pfleger collection fits into this range, with an emphasis on colours such as cream, oatmeal and white, cut for soft comfort, with dark blue and black accessories. Their new shop in Berlin presents this collection in an appropriate setting.

The shop footprint is quite small, but covers over two floors and basement. The walls are painted in a soft white, with white treads on the curving double flight of stairs between the two levels. The floors use plain coir carpeting (a natural material, untreated), shelving is in white, and furniture is in plain pale wood. This neutral/natural environment needed careful lighting to animate it without dramatising it. Lighting designer, Alexander Kovacs worked alongside the interior architect to achieve this softly dappled interior in which the total 'look' – of clothing, interior and lighting – is harmonised perfectly.

Looking up the stairwell *(above)* shows the
regimented patterning of fittings that create a
dappled effect on the shelving and merchandise in
the interior *(facing, below)*.
The main stairwell is emphasized by combined
downlighters/uplighters fixed to the stair walls *(facing,
above)*.

The first element to be considered was the shop window. Because of the small footprint of the shop a separate window was not an option, so the whole embrasure, floor to ceiling, was planar glazed in two panels, behind which stand mannequins and low display shelves, with the rest of the interior visible behind. These are lit by low-voltage tungsten halogen fittings, fitted with barndoors, and by the recessed downlighters fitted across the whole ceiling. These downlighters use lamps with narrow beams in groups of four to create soft pools of light on the flooring. In the main areas these are supplemented by tracked spotlights, again with barndoors, to give extra light to shelves or garment racks.

The stairwell was another feature that needed emphasis. Thus, the stairwell is lit by paired uplighters/downlighters mounted on the walls, as well as by spotlights from the track on the first floor. The result of this careful lighting plan is the creation of a cool but welcoming interior, in tune with the design mission of the collection, and yet flexible enough, with tracked lights, to be adaptable to new ranges of clothing and accessories.

Joseph Shop, Fashion, London, UK.

> *'...the main challenge was the strong level of daylight in the shop.'*

Joseph Ettedgui's clothes say style, authority, edge. Exquisitely cut black suits and boldly branded accessories in the late 1980s have given way to softer colours in the 1990s, without losing any of their boldness or attack. The Prague-born architect Eva Jiricna has worked on the design of many Joseph stores, including the one in Brompton Cross in London. This store comprises a ground floor and basement linked by a suspended steel staircase.

'Joseph asked me for an Italian palazzo', Jiricna explains, 'but the main challenge was the strong level of daylight in the shop. It is not often realised how levels of daylight can affect the lighting design of an interior. A second problem was the low ceiling height in the basement, which reduced our choices for lighting it.' Rather than trying to block out the daylight, the design accepts and enhances it. 'The glass treads on the staircase were intended to let the light through, and we deliberately concealed the light fittings because the effect of light – artificial and natural – was what was important.' The light is used to cover and give dimension to the spaces.

In the basement area the main lighting is from fluorescents behind recessed flush ceiling fittings. Tracked fittings would both have been inappropriate and posed height problems.

The scalloped detailing on the columns is mirrored by the curved pools of downlighting against the walls *(facing)*.

On the upper floor the main light source is from recessed tungsten halogen ceiling spots, which pick out the detailing of the stair construction and the elegant modern furniture, including a table by Ron Arad. 'We put fluting around the supporting columns on the ground floor to vary the appearance of light', Jiricna comments, 'and matched this with wallwashing pools of light from recessed fittings along the side and end walls'. The fittings are not simply recessed into the ceiling, but deeper, into a slot within it, so that no part of the lamp or fitting is visible, except from directly beneath.

All the surfaces, especially the floors and the stairs, are highly reflective. Thus, they require fewer fittings and a lower lumen package to be lit effectively. Excessive lighting would risk glare or distraction from the goods on display *(left)*.

Jiricna's design combines austerity of purpose with formal elegance. The choice of finishes and materials, the discreet use of decoration and the handling of lighting show both creativity and control. She has created a very modern palazzo using wholly contemporary materials. But, as the scholar Howard Burns has pointed out, the architecture of Palladio himself 'shows how adherence to method and system can facilitate the creation of totally new solutions.'

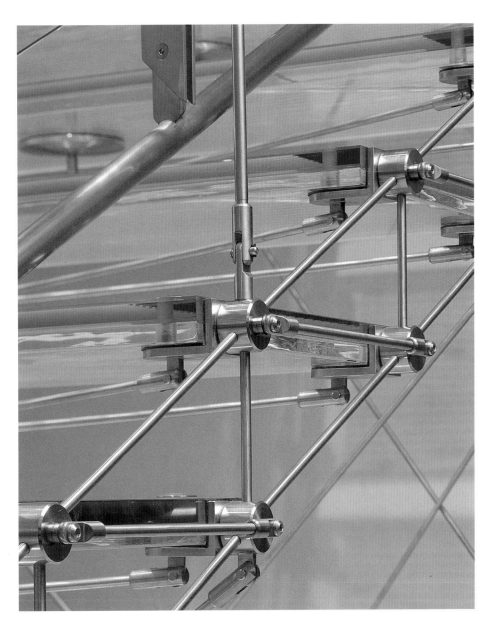

The functional steel engineering of the staircase makes a strong aesthetic and technical appeal. The lighting gives a deliberate sparkle to the skeletal elements of the staircase *(above)*.

In the basement, the glass treads on the steel stairs allow light through. The recessed ceiling lighting simulates cool daylight with linear fluorescent lamps *(left)*. On the main floor *(above)* note the use of marble floors for reflectivity, and the use of contemporary furniture for styling.

Dr Martens Department Store, Clothing and Shoes, Covent Garden, London, UK.

'Part retro, part redneck.'

The social movement of clothing in the last 30 years has run counter to earlier patterns of styles cascading down the social order from the aristocracy. Now clothes move up: the classic example of this are denim jeans, originally working-class workwear, now universally acceptable and graced by *haute couture* names. The Dr Martens boot is similar, its famous Airsole first devised as a protection against acid, petrol and other industrial risk products. The shoe was then taken up as a social symbol by punk rockers, and has now moved on to a global fashion stage where its wearer's feet are unlikely to test the sole's special properties at all.

The industrial origins of the shoe were behind the design briefing for the Dr Martens flagship store in a five-storey building in London's Covent Garden. This opened in 1994 to show not only footwear but also the new Dr Martens men's clothing range. The London-based designers DIN Associates were invited to handle the interior decoration and lighting.

'We wanted to create a strong feeling of reality, to show what everything was made of', John Harvey of DIN Associates points out. 'We stripped the walls back to bare brick, used metal lattice oil-rig decking on the walls as supports for shelving, and left the aluminium ventilation trunking deliberately exposed. This created an industrial look, to which we added strong colour details through the furniture, fittings and display items.'

The entrance lobby, deliberately rebuilt as a double height space *(facing)*. The shelving is attached to oil-rig decking bolted to the stripped walls, part of the in-your-face design approach.

SECOND FLOOR
VIEW TO FRONT AREA

As to the lighting to be used, the obvious starting point was industrial lighting systems, the simple but robust bulkhead lights and pendant downlighters found in factories and warehouses. 'I'd found an old desk light on an expanding frame in Barcelona,' Harvey explains, 'which had just the right feel, part retro, part redneck.' This and a plain-spun aluminium half-sphere downlighter provided the visual cues for what became a series of special lights developed for the unique use of the store. The original fittings only used standard GLS lamps, quite inadequate for modern retailing, so DIN Associates carried out a series of tests, finally opting for warmer white SON lamps (rather than cooler metal halide lamps), with a variety of openwork grilles or etched glass covers for the downlighters, and simple dichroics, rather than tungsten halogen, with metal reflectors for the track-mounted luminaires on extendible arms.

LIGHTING

HIGH BAY
LIGHTING

STAIRWELL AND
LIFT LIGHTING

SERVERY LIGHTING

AMBIENT RETAIL
LIGHTING

STAIRS DISPLAY
LIGHTING

FITTING ROOMS

SPOTLIGHT FOR SMALL STORE
DIRECTORIES

Design sketch showing use of modified industrial
fittings *(facing, above left)*.
  Presentation board showing the range of fittings
proposed, a range of special designs using the visual
language of traditional industrial fittings with
modern lamps *(facing, below left)*.
The main lighting is from modified high-bay dome
fittings mounted on descending extendible arms
(above).

Even in areas such as toilets, care is taken with the lighting, here provided by a recessed fluorescent strip *(above)*.
Standard industrial bulkhead lights are used on the stairs, with etched glass replacing the clear original cover *(right)*.
Strong colours placed strategically behind display areas increase the visual buzz *(facing)*.

The main entrance *(above, left)*. Note the recessed fluorescent ceiling strip acting as an edge marker.

In the main sales area *(below, left)* receding planes of lighting (parallel lines of recessed downlighters, angled wallwashers, and toplights to display racks) illuminate the goods on display.

Mitsumine, Menswear, Kashiwa, Japan.

'Security and confidence are the watchwords behind the design solution.'

The Mitsumine men's store in Kashiwa, Japan, sells a range of formal and informal menswear in Western styles. These are classic clothes in modern styles and colours, and the interior design reflects this. Wooden parquet flooring, off-white columns, and showcases in dark varnished wood echo the atmosphere of a formal but not too traditional club. Security and confidence are the watchwords behind the design solution. The store is in a mall, with an open entrance area. Here the aim of the design is to draw the customer in, and the lighting is used in layers of fittings, from the central large downlighters on the ceiling spine through two lines of recessed smaller downlighters, and then wallwashers under the cabinet frames. This creates an orderly and inviting progression towards the merchandise. To highlight the goods on display, the showcases are backlit with fluorescents behind tinted glass. The design solution has the welcoming formality of a traditional invitation card. Mitsumine – and their customers – are at home.

In the main sales area *(left)*, display cabinets are backlit with fluorescent light.

The cast iron columns with their elaborate capitals dominate the centre of the high-ceilinged space, flanked by a sinuous line of black leather bench seating *(left)*.

The staircase is a modern addition, but could have been appropriate to the original function of the building. It creates a sculptural bridge between the past and the present use of the space *(above)*.

Joan and David Shoe Store,

SoHo, New York, USA.

'The design is inserted into the space rather than seeking to dominate it.'

Joan and David sell clothes, shoes and accessories. Their first store, on one floor of an old five-storey industrial building at the bottom end of New York's Fifth Avenue, combined existing features, such as the 19th-century cast iron columns and the open plinthed ceiling, with new fittings and decor devised by the architect Eva Jiricna. This once unfashionable part of New York became the area to live and shop in during the late 1980s, with its galleries, lofts and markets made famous by Tamara Janowitz's *Slaves of New York,* among other books. The special character of the area and its architecture is respected in Jiricna's sensitive handling of the interior. 'I did not want to interfere with the quality of the existing space,' Jiricna points out. 'So the design is inserted into the space rather than seeking to dominate it.'

In the single-floor sales area, the lighting, from tracked tungsten halogen light fittings, is set around and above the display areas on the walls, leaving the architecture to make its own statement in the centre.

The main feature introduced into the space is the staircase, which connects only to a storeroom. 'I see it as a sculptural statement,' Jiricna says. The staircase, set against one wall, provides an alternative visual focus to the rows of columns. The openwork steel treads, and the use of a wire system for the banisters, reflect in a wholly contemporary way the industrial origins of the building.

The Joan and David shop shows how the careful use of lighting can both display what is on offer and enhance the qualities of a strongly architectural setting. The background colours are deliberately minimal (silver, white and black), while the choice of materials (steel and wire) and the exposed lighting system chime in with the functional origins of the building.

The shelving system is deliberately minimal, using simple white metal shelves supported by industrial tie-bars. These are backlit using the reflective wall panels to spread the light *(left)*.

The perforated treads of the staircase allow light to wash through into the space below *(facing)*.

Staying modern in a fashion store: note the polished steel table by Israeli designer-sculptor Ron Arad, and the grouping of downlighters above it to make it centre stage. The clothes themselves are arranged in alcoves (rather than on a rack) and generously lit by downlighters within the alcove frames *(above)*.

Signposting a couture designer's section in a larger store: note the spotlight on the mannequin under the Ralph Lauren name *(left)*.

Approaches to Desire

Different shops offer different levels of welcome: cane chairs for a leisured consideration of luxury fashion, visual drama through deliberate underlighting, careful choice of translucent materials to aid display. The vocabulary of concepts at the lighting designer's disposal allows the subtlest nuances of marketing to be enshrined in the lighting solution. The examples on these pages show a number of 'tricks of the trade'. But the underlying theme is the acceptance of the need for a lighting solution to be a complex system, which has the opportunity to develop and evolve built into it from the start.

Contrasting lamp colours (through the use of cold linear fluorescents) mark out different areas in a Japanese accessories shop. The surface materials have low reflectances except for the pale wood of the display tables *(above)*.

In the Hugo Boss shop in London, the deep ceiling slots for the recessed track spotlights become a visual feature, enhanced by the choice of black luminaires *(left)*.

All parts of an interior design have to make a whole: this is equally true of lighting to be successful. A successful solution will meet the immediate lighting needs of the space, be simple to operate (in terms both of relamping and switching different parts of the system on and off) and be able to be adapted to meet new display requirements for new ranges of goods. This is particularly true of the fashion business, with its seasonal changes of colours, but applies in the total retail sector as well. A track system, whether ceiling-mounted or suspended, is the obvious example of how to create these opportunities for change, and track systems are widely used for this reason. But fixed-position lighting can also have inbuilt variability through the careful planning of circuits to allow for different combinations of sources to be lit together or in groups, and through choosing fittings that can accept different lamps, with alternative outputs, beam widths or light colours, and which can be modified by additional elements such as diffuser rings, barndoors and filters. The designer's job is to make the client's task of selling easier, and this is often best achieved by allowing for the future while meeting the demands of the present.

Curving frosted glass shelving at the Bergdorf Goodman store on New York's Fifth Avenue allows diffused light from wallwashing downlighters to illuminate the whole collection evenly, layer by layer *(right)*.

Armani men's suits on display. Lighting colours and levels need to be carefully controlled to bring out the subtle range of fabric colours and textures. There are three sources of light here: the daylight from the back of the site, accent lighting from tracked spotlights and a general sparkle from tungsten halogen downlighters fitted with conical frosted glass diffusers *(facing)*.

Food retailing is becoming an increasingly important retail sector, through new hypermarkets and superstores, as well as in conventional supermarkets. At the opposite end of the size spectrum, specialist food outlets offering gourmet rarities, imported cheeses or fine wines from around the world are also competing for customer's attention.

The technical considerations that are important in designing a system for lighting food are twofold.

Firstly, attention needs to be paid to the question of heat generated by lamps and fittings, particularly if fresh food is involved. The food business today is extremely conscious of health and safety aspects, and the lighting designer should understand these concerns. Secondly – and this is particularly true of fresh food – the appearance of what is on display has a crucial effect on decisions about purchases. Vegetables should look crisp and fresh, fruit bright and cool. Colour rendition is thus a key factor in the selection of lamps.

General interior shot of Carluccio's Food Shop in London's Covent Garden *(facing)*.

Carluccio's Food Shop,
Covent Garden, London, UK.

'Expresso coffee, flavoured coffee, and Florentine drinking chocolate...the best of all things Italian.'

Italian food has become seriously gourmet in recent years, moving away from industrial pasta with dubious sauces to original, regional dishes. Antonio Carluccio's Neal Street Restaurant has long been one of the best Italian restaurants in London, and some years ago he and his wife Priscilla opened a delicatessen to sell imported regional specialities such as mushrooms (a favourite item in his own recipes), oils and cheeses.

The Food Shop is set in a small, rectangular space next door to the Neal Street Restaurant in London's Covent Garden. A deceptively small space, for in addition to imported products Carluccio's also creates and sells its own range of products, such as expresso and flavoured coffees, and Florentine drinking chocolate.

The setting is kept deliberately simple, with a traditional marble-topped counter to one side and adjustable openwork metal racking supporting the goods on display on the other. The space is lit by natural daylight both from the fully glazed entrance and from a window and glazed door in the back wall.

Wallwashers over perimeter shelving and dimmed downlighters over the counter display create an informal and intimate ambience *(facing)*.

The lighting solution was both direct and simple. There are three rows of recessed ceiling downlighters with adjustable angled 50-watt dichroic tungsten halogen lamps. These provide a general wallwashing downlight giving a good horizontal illumination on shelving and display, with a concentrated bank of downlighters over the marble counter. Floor and walls are in pale whites, with a highly-reflective white, stepped ceiling to increase the sense of space and activity.

A supplementary, or alternative source of light, comes from a set of wallwashing uplighters along the unshelved wall. The combination of these two systems and the relatively high level of daylight provides a variety of opportunities to light the space in different ways.

At night a double illuminated neon sign sits over the fully glazed entrance *(left)*.

The glazed rear door provides both a colour reference and adds to the overall light level *(facing)*.

Despite its small footprint, the Food Shop is very efficiently organised. Inside, industrial racked shelving allows different products to be displayed simply, flexibly and directly. The combination of all three systems – daylight, wallwashers and downlighters – gives sheen and sparkle to the products displayed. The tungsten point sources are used to create both shadows and highlights *(right)*.

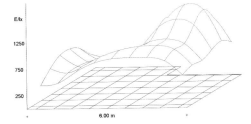

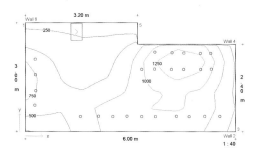

Three-dimensional computer rendering of light levels within the shop *(above, left)*. Computer rendering of light levels at the working plane *(left and below, left)*.

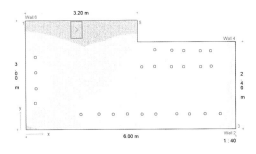

Calligraphed statements about whisky lit by narrow
and broad spotlights in the exhibition area at the
Glenlivet Distillery Visitor's Centre *(above)*.

Glenlivet Distillery Visitor's Centre and Retail Site, Glenlivet, Scotland, UK.

' The Glenlivet is unique. It has defied all attempts at imitations.'

Marketing people talk often about the 'brand experience', and for malt whisky distillers this is an ongoing challenge. The quiddity of a malt whisky brand depends on its individual nature – its creation from local water, peat and barley – but it is sold on competitive shelving with not only other malts, but blended whiskies as well. Creating a strong brand identity is therefore paramount, and part of this is establishing in the consumer's mind the special properties of the drink. Visitors' centres at distilleries, where the brand, its values and content can be appreciated in person, play a useful role here.

The challenge faced by award-winning Scottish design group Northcross to reanimate the Glenlivet Visitor's Centre – or brand experience – was therefore considerable. They responded both enthusiastically and subtly, creating a movement framework for visitors that not only contained the traditional distillery tour but an exhibition site, a cafeteria, meeting rooms including a ceilidh area, and a retail outlet.

Displays of racked bottles lit at an angle for
emphasis *(above)*.

The retail sales desk: note the natural materials
emphasised by warm light sources and controlled
beams of light *(left)*.

The exhibition area contains material on the history of smuggling and the later development of legal distilling, an electronically activated model of the distilling process set on an interactive table and an audiovisual presentation of 'The Elemental Spirit' to emphasise the natural elements and processes that go into making whisky.

The lighting in the exhibition area and the retail section was deliberately theatrical – tracked spots often fitted with gobos, special fittings in wrought iron holders – intended to involve and engage the viewer in the display. In the ceilidh area there are pastiche special fittings decorated with stags' antlers to reinforce the central message that Glenlivet is an original, sophisticated, traditional whisky made from natural ingredients.

The interactive table: a hand passing over the items on the display triggers overhead spotlights and sounds *(above)*.

Lighting mounted on the roof beams in the exhibition entrance *(right)*.

A specially designed light fitting in the shape of a stag's head *(far right)*.

Carlton Hobbs, Antique Dealers, London, UK.

'Rich and precious objects in a courtly ambience.'

Dealers in fine art and antiques also have the challenge of lighting continuously changing stocks of work on display. The London firm of Carlton Hobbs chose a sophisticated array of spotlights to give them the necessary versatility. The spotlights are mounted on a Lytespan twin circuit track and use a range of seven different lamps and fittings combinations to highlight the various surface textures on display: gilt and crystal, lacquer and velvet, hardwood and porcelain. Lens systems on the spots allow different beam widths and focus points, and multi-track switching to highlight different areas. The ceilings and upper walls, though not high, are deliberately underlit, allowing the lighting to create a rich and courtly ambience at ground level, without the fittings interfering visually with the work on display. The design intention is to create a space reminiscent of a museum or country house, with different elements – paintings, furniture and *objets d'art* – grouped and lit together, as if in a *wunderkammer* or cabinet of curiosities. Only a sophisticated and well organised lighting system could achieve this, and respond to changing demands.

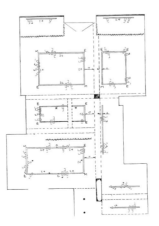

The showroom *(see plan, right)* is designed to be reminiscent of a museum or country house, replete with paintings, furniture and *objets d'art (facing)*.

In this window display uplighting on the mirror is combined with downlighting on the table to isolate and define the objects *(above)*.

Dark ceilings with carefully positioned spotlights create an almost theatrical atmosphere *(above)*, while a framing head is used to isolate a painting *(right)*.

Stockman Mannequins,

London, UK.

'A place where professionals meet professionals.'

Mannequins without clothes are either a design statement or an admission that the window dresser has gone out to lunch, but at Stockman Mannequins they are the stock in trade. The company supplies all types of display busts and equipment to the fashion trade worldwide. Their showroom presents their own collections, and those produced by Pucci International of New York, to professional buyers.

The showroom measures almost 14 metres by 10 metres, with a graceful curving wall opposite the entrance. The space is mainly lit by artificial light, and is broken up by four structural columns. As it is a space where professionals meet professionals, the decor is simple. Off-white walls and ceiling, with a stripped and varnished wooden floor. The bust forms are displayed in groups, lit by spotlights mounted on black tracks running between the wall and the pillars.

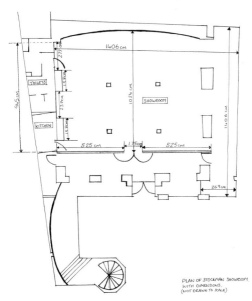

PLAN OF STOCKMAN SHOWROOM
WITH DIMENSIONS.
(NOT DRAWN TO SCALE)

Highlights from tracked spotlights create pools of light over the display busts (facing).

The plan (right) shows the position of the lighting track.

Variable light levels are important when apparently equal levels of illuminance are required for adjacent objects of different colour and tonal depth. More light is required on black objects than white, for example, because of their different reflectances *(above)*.

The track system *(facing)* allows the light to follow the changes to the display arrangements – an endlessly flexible solution while the ambient light allows detailed, pensive study *(left)*.

The result is a space which is both dramatic and practical. It has a flexible system with the ability to move spots to different positions. The spots use 100-watt 12-volt filament capsule lamps with different beam widths created by clip-on reflectors. The tracked spots allow for variations in both lighting levels (a three-step dimmer allows variations of intensity) and effects (since alternative lenses, filters and barndoors can be fitted to the spots). And the light quality from the spots is very similar to lighting levels and types used in shop windows, so that the customers can appreciate the product in what will be its fully clothed surroundings. The success of the space is seen in its use not only as a showroom but as a promotional space for parties and launches, and as a photographic studio.

National Film Theatre Bookshop,

London, UK.

'Lights, camera... books.'

The lighting of bookshops presents an interesting challenge: books are a category of merchandise examined in detail before purchase, and bookshops a natural space for customer browsing. The lighting needs to be fairly strong, to allow author's names and titles, blurbs and covers, to be read easily. Specialist bookshops in addition need to carry a large range of titles, which customers can select and inspect themselves. The wall shelving often forms the main display area, which in turn needs good, even lighting from top to bottom so that titles can be read at an angle to the main line of vision. Within these constraints, finding a visually interesting solution can be difficult. The device used at the National Film Theatre bookshop in London was a mirrored ceiling. This increases the light levels by reflection (without the risk of glare or heat that adding more fittings might create). It also adds interest to a visually tight space, without causing any confusion.

The light comes from independent spotlights mounted on fixed vertical rods between the mirrored ceiling tiles. These ensure an even wash of light over all the shelving and the central displays. The relatively generous vertical spacing of the shelves (in part to accommodate different book sizes) also enables the lighting solution to work well.

Face-out or spine-out, books need to be browsed. Overall lighting from fixed spotlights on ceiling pendants creates even illumination *(facing)*.

Lighting Opportunities

Every retail environment deserves the care and attention of a lighting designer. Even a setting as functional as the racks of a self-service wholesaler can be enlivened, and made more workable by good lighting. The opportunities for innovation and experiment are endless, fuelled by the increasing diversity of the retail market.

Continuous fluorescents positioned to wash light over the shelving in a small French supermarket *(above)*.

A simple wooden floor and daylight add to the open air feel of this Wolky shoe shop in The Netherlands. Paired spotlights suspended from tracks in the high ceiling supplement the natural light *(left)*.

The first European department store was the Bon Marché in Paris, which opened its splendid bronze doors in the mid-19th century. The concept of offering a complete range of household goods, clothes, furniture and accessories was born. The idea spread with many department stores opening in London, New York, and across Europe by the end of the 19th century.

The shoe department at Harrods in London, designed by DIN Associates, uses ceiling downlighters with glass diffusers, fluorescent striplighting along the ceiling rim and uplighters in the wall alcoves *(facing)*.

Department store shopping has always been seen as luxury shopping. Ornate fittings, marbled floors and elegant presentations have all been at the heart of this part of retail culture. And despite competition from boutique culture in the 1960s, malls in the 1970s, megabranding in the 1980s and virtual shopping in the 1990s, department stores have managed to adapt and survive, sometimes becoming themselves brands in the process. Names such as Saks Fifth Avenue, Galeries Lafayette, Harrods or Bloomingdales have an equity value that is very considerable.

Lighting played an interesting part in the development of the department store. The first stores relied on newfangled inventions such as gas or even electric light, supplemented by daylight from generous windows and often from skylights or glazed domes, such as that in the original Bon Marché. The later development of reliable and sophisticated electric lamps allowed the interiors to be entirely lit by

artificial light. The ground floor windows became enclosed showcases, the upper windows were blacked out, or often became storage areas, as the retail selling environment turned inwards to enclose the customer entirely. The customers literally entered a world apart, not only in terms of the luxury of the surroundings, but also by being isolated from visual contact with the exterior world – an early example of what a marketing expert would today call a 'brand experience'.

The wall-mounted display boxes in Harrods shoe department are backlit with compact fluorescents behind a translucent screen, with the ambient light coming from the ceiling fittings *(above and left)*.

The wall alcoves at Harrods shoe department are lit in daylight colours, to simulate window embrasures, although their positions do not align exactly with the original window spaces *(facing, far right)*.

Detail of the display cases *(facing, right)*.

Galerías Pacífico, Shopping Centre, Buenos Aires, Argentina.

'Retaining the original splendour through renovation and relighting.'

The entrance area with its columns lit by crossed spotlights *(facing)*.

One of the most famous arcades – the precursors of department stores and malls – in Europe is the 19th-century Galleria in Milan. It flanks the cathedral square in the centre of the city, and its plan is a cruciform shape, lit from above by daylight through a glazed central dome and roofs. Its mixture of cafés and exclusive shops make it the social heart of fashionable Milan. In Buenos Aires in Argentina, the Galerías Pacífico is also the place to shop, in an equally elegant and also cruciform setting, though on two levels compared to the Galleria's one. When it was decided to renovate the 19th-century building, the architects Juan Carlos Lopez and Associates called on Theo Kondos from New York to help with the lighting design.

For the entrance to the arcade, with its imposing Composite columns, crossed uplighting floods were used. On the main level a pattern of lights was hung from the glazing frame of the roof, while supplementary crosslighting on the upper levels highlights the architectural details of the façade. The central glazed dome was crowned with lights on the glazing bars, with blue-tinted concealed fluorescents in the squinches. In the lower-level crossing point the original murals were refurbished, so this became a focal point for the centre. Museum-quality lighting was used to protect the murals from fading through overexposure. The lit fountain also animates the central court, as well as providing a meeting area. Throughout, a careful combination of daylight and artificial light creates a welcoming and relaxed ambience.

The restored mural and lit fountain in the lower court *(above)*.

One of the four arms of the cross, showing the lighting under the glazed roof and the crosslighting at the upper level *(left)*.

The glazed dome at the centre of the mall: note supplementary coloured lighting in the squinches *(facing)*.

Saks Fifth Avenue, Department Store,

Beverly Hills, California, USA.

'By creating highs and lows in the lighting, we were able to create a dramatic, up-scale ambience.'

Saks Fifth Avenue is normally associated with that grandest of Manhattan streets, but Saks Fifth Avenue is also a brand name, with branches in other major American cities, including Beverly Hills in California. The opportunity to expand their store site, and to renovate an interior that dated from 1985, led Saks to approach the New York interior design and marketing experts FRCH Design Worldwide.

The total square footage is 325,000. The original five-floor building is 210,000 square feet, with the new wing approximately 15,000 square feet and the former Magnin building across the street with the Menswear Department, 100,000 square feet.

The club setting of the new menswear shop *(facing)*.

The exterior of the main building *(left)*.

The fashion department *(facing and above)* uses fixed display racks lit by a combination of double fluorescents, recessed adjustable PAR halogen mains-voltage spotlights and low-voltage halogen spotlights and T5 miniature fluorescents in the display units.

FRCH Design Worldwide, formed by the merger of design groups SDI and NTI in 1990, was able to bring a range of specialist talents to the task. As well as lighting design, their responsibilities included interior design, colour and materials, merchandising and fixture design.

The design brief emphasised Saks' market position as up-scale, classic and stylish. In addition, FRCH was challenged to retain as much of the original architecture as possible for budgetary reasons while making sure that the existing Saks building, the new addition and the converted Magnin's building would all seem like the same store to the consumer.

A key part of the renovation of the main store was to open up the original ground floor area into a single unit, thus opting for a grander opening statement. White marble floors with gold veining, silver-leaf accents and a warm colour scheme helped to establish a luxurious atmosphere.

The lighting was a combination of two-by-two fluorescents; fully-recessed, adjustable, PAR halogen accents; low-voltage halogen accents in the casework; and the new T5 fluorescents.

Cynthia Turner, Vice President & Director of Lighting Design, FRCH Design Worldwide, was responsible for the lighting design along with architects Bridges and Lavin and Saks Fifth Avenue's in-house design team. She comments: 'In most of its stores, Saks prefers to use a lot of incandescent lighting, but for the Beverly Hills store we had to take into consideration California's more stringent energy codes. Instead of relying heavily on incandescents (which are not energy-efficient), we used more warm-coloured compact fluorescents. Also, we were careful to vary the light levels in the store by focusing the lights on the merchandise, rather than the architecture, as much as possible. By creating highs and lows in the lighting, we were able to create a dramatic, up-scale ambience and avoid the flatness found in discount stores'.

'In the original building the challenge was to incorporate the new lighting design into the existing architecture, mouldings and plaster work and at the same time, make sure that the lighting quality and colour were consistent throughout the store, including the new wing and the renovated Magnin building. All of this had to be accomplished within a limited budget'.

For the new men's shop on the adjoining site, design co-ordination was the key. Here, a golden colour range and rich wood finish were chosen, linking the prototype of the gentleman's club with the elegance of nearby Hollywood's golden age.

The menswear store uses traditional pendant ceiling lights supplemented by two rows of tracked mains-voltage spotlights *(above)*.

Galeries Lafayette, Berlin, Germany.

‘The only possible context for a design is its epoch.’

The fall of the Berlin Wall saw the start of a building spree in the former East German capital, matched by a whole range of new commercial and economic ventures in the former European Communist bloc. There is now a Rolls Royce dealership in Moscow, which someone said was proof at last of Lenin's dictum that 'one day all comrades would travel first class'.

Galeries Lafayette, the famous French department store, decided to create a flagship in Berlin in a new building designed after a competition by the world-renowned architect, Jean Nouvel. He took the bold step of inverting the traditional glass dome that skylined so many famous 19th-century stores, and turned the concept into one immense glass funnel that drops into the heart of the building from roof level, while another rises from the floor. The Galeries Lafayette store is at the heart of these, so that all of the sales floors can be taken in at a glance, while the central cone is also continuously lit by coloured anamorphic projections, creating a whirling pattern of luminous dots, a metaphor for the flow of information and sensation that is central, in Nouvel's view, to the modern urban experience.

The sculptural and imposing exterior of the Galeries Lafayette building on the Friederichstrasse in Berlin *(facing)*.

Elavation of the Galeries Lafayette building *(right)*.

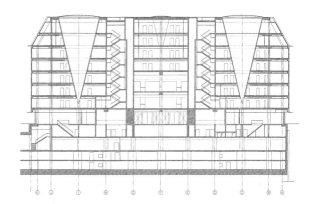

Nouvel's radical approach to lighting the Lafayette cones involves treating light as part of the information system of the building, rather than as a functional element. On the shop floors themselves, conventional lighting systems are used to display goods and products, with the curved walls of the cone puncturing every floor with its flashing message of change in activity. This is not merely design opportunism - the idea of making a grand statement - rather his intention was, in his own words, 'to bring the urban reality into the interior of the building through the metaphor of light.'

The interior of the main cone: it forms a well of light cutting through the floor levels which are conventionally lit, but it is also a projection surface for anamorphic images and patterns of coloured dots and lines *(facing)*. Constructional details of the cone *(right and above, right)*.

Various different views showing the constructional details of the cone *(left and above)*.

SUPER, HYPER, MEGA

The shopping mall, once famously described by the Vice President of Columbia University as an 'icon of the modern American lifestyle', has undergone many developments and changes since the concept was introduced in the early 1970s. With the changes in consumer attitudes of the 1990s, the need to reinterpret the mall concept has become more pressing as the desire emerged to redesign malls built some 20 or 30 years ago.

A typical mall arrangement with discreet ceiling lighting from spots mounted on vertical tracks. This allows the shop fronts and logo space to make their own statements. Note the use of soft-coloured floor tiles to warm the otherwise bleak ceilings and pillars *(facing)*.

The first malls tended to use the central metaphor of a covered street, harking back to the arcades that graced many 19th-century European cities. Thus, parallel rows of shops would be roofed over in glass and a formal entrance portico created at the front. This is still a valid solution on a small site, provided the quality of the interior architecture of the mall itself is sufficient to make the visitor feel welcome. But with the decision to incorporate not only shops but cinemas, bars and restaurants into malls, they have become almost 24-hour centres, where a family or a group of friends can not only shop, but also eat and be entertained. Thus, the mall no longer serves as a backdrop to retail activities, but has acquired a positive function offering a complete 'experience'. This has important consequences for the design

The entrance to a 'parfumier's' shop in a mall. The use of coloured fluorescents in the ceiling recess marks off the open entrance from the main mall in Hong Kong *(above)*.

approach: the modern mall has to have its own design personality which will bring together, but not dilute, the different elements it contains – a principle that applies just as strongly to the lighting solution.

The mall as street metaphor can still work well on a small scale, such as the Wellgate Centre in Dundee, Scotland. Here a formal glass ceiling on bold steel arches covers three levels of shopping and a basement. The aim here is to create an atmosphere reminiscent of famous historical department stores such as the GUM in Moscow, or the Bon Marché in Paris.

The central well is left deliberately clear at Wellgate and is lit by daylight supplemented by luminaires on the balustrades of the two upper floors, uplighters on the ground floor columns and wall-washing recessed downlighters beneath the overhang of the first floor *(right)*.

At ground level in the Wellgate Centre there are rows of recessed downlighters in the ceilings using Metalarc lamps in Equinox fittings, as well as uplighters on the interior faces of the supporting square columns *(above)*.

The integration of luminaires into the overall design is an important factor. At Wellgate and at the Merseyway Centre in Stockport, special fittings were designed and produced to give each environment a distinctive visual quality. At Wellgate the lamps, in frosted white and green glass bowls, were built into columns rising from the wrought iron balustrades with the same design used for the wall-mounted uplighters on the ground floor columns. At the Merseyway, paired wall-mounted uplighters with brushed aluminium bars and holders were crowned by frosted-glass diffuser rings.

The specially designed uplighters at the Merseyway Centre, created by lighting designer Barry Wilde. The uplighters are placed above the row of shops to avoid visual interference with shopfront details *(above and left)*.

Times Square Department Store, Hong Kong.

Times Square in Hong Kong is one of the premier department stores in the city. It creates its own sense of being special through the internal organisation of space, for the ground floor forms the basis of a hollowed-out ziggurat. Above the visitor's head a series of eight floors rises up, each one overlapping the next one until the ceiling is reached. The profile of each floor is picked out with cold cathode lamps programmed to change colour from blue to purple to gold. On each floor, the pedestrian access area is lit by recessed downlighters with compact fluorescent lamps mounted at one-metre intervals.

Looking across the levels of floors *(facing)*.

Looking up at the array of floors within the store: note alternative colour and light combinations that allow variety within the overall continuity. This is achieved by tinted cold cathode lights outlining the soffit edges of the ceiling panels *(right)*.

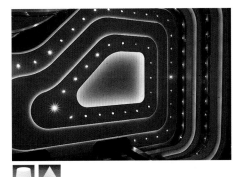
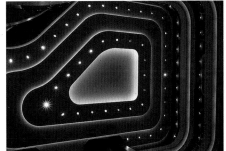

The Shops at Harrah's Casino, Las Vegas, USA.

'The stores are filled with boldly-themed environmental graphics, unusual materials and strong colours. We used the lighting to bring out the drama of these textures and colours.'

Primary colours and gobo stars at Harrah's Carnaval Corner *(facing)*.

Las Vegas is, we have been told, a place we can learn from. A city devoted entirely to the commerce of pleasure, it is being continuously refashioned as new features, theme parks, malls and hotels are added around its core of casinos. When deciding to extend and upgrade Harrah's Casino, the owners decided at the same time to rework the retail area, and invited FRCH Design Worldwide to work with them.

The collaborative process between FRCH and Harrah's began with image positioning and concept development. Working closely with the client, FRCH produced a complete design and documentation package, including architectural documents, interior and storefront design, theming, graphic communication, fixturing, propping and merchandising, for a series of five stores. The new retail concepts include a party store, an art gallery, a sundries shop and stores selling a broad assortment of branded merchandise based on Harrah's live shows and the casino experience. The shops, which range in size from 1,000 to 5,000 square feet, are strategically located to draw traffic from within the resort, as well as from tourists and residents on the city's main thoroughfare.

Although each store has its own distinct design, they all reinforce the company's theme: 'Harrah's is about great gaming.' By capturing the excitement of the casino, the retail environments are able to capitalise on Harrah's brand equity and become part of the overall gaming experience. Bright colours, bold theming, striking shapes and theatrical lighting establish an upbeat, celebratory tone that's in keeping with the casino's party mood. Now the first stores have opened, FRCH is working with Harrah's to extend its retail identity beyond the casino with a new store in the Las Vegas airport.

147

The lighting design had to reinforce the upbeat, festive tone of the new shops in Harrah's Casino. The lighting also had to contribute to the up-scale, speciality-shop ambience which Harrah's wanted to appeal to the 'high rollers' who visit the casino to gamble and enjoy the live entertainment. In addition, the lighting design had to be fairly flexible since a lot of the merchandise that would be sold in the new stores was still being developed while the retail environments were being designed.

'Accordingly,' says Cynthia Turner, Vice President and Director of Lighting Design at FRCH, 'we used a variety of theatrical lighting effects, such as colour gels and framing protectors with gobos. We also mixed theatrical fixtures (PAR lamps and theatrical tracks) with architectural fixtures modified to look like theatrical fixtures (gimbal rings holding PAR lamps). In some areas we used fluorescent downlighters for ambient lighting.'

At Harrah's, on-stage devices like the exposed lamps round the mirror *(facing)* and the director's chair *(right)* set the scene.

At the Carnaval Corner *(facing)* tracked spotlights can be adjusted to follow mobile food displays, while at the Jackpot Clothing Store *(above)* the star gobo motif dances over the merchandise.

'The stores are filled with boldly-themed environmental graphics, unusual materials and strong colours. We used the lighting to bring out the drama of these textures and colours. To maintain a strong contrast we had to be careful to keep the accent lighting away from the video monitors which are in some of the stores. The lighting was designed to produce highs and lows for these reasons and also to reinforce the up-scale ambience.'

The architect designed the two main towers flanking the entrance as a steel skeleton. Lighting the towers at night creates a dramatic, space-age striped three-dimensional form, with a dark sky as the backdrop *(above)*.

CentrO Retail Centre, Oberhausen, Germany.

'Transforming the old industrial landscape.'

If retail development is symbolic of changes in lifestyles, CentrO, the largest retail site in Europe, can stand as a model. Built on the site of a former steelworks, it represents the change from heavy industry to services, and with its themed restaurants and linked leisure facilities is a fine example of contemporary marketing. The design was masterminded by RTKL, the international planning, design and architectural firm. CentrO, Oberhausen, is situated on the site of former Thyssen steel mills outside Dusseldorf; the scheme is a symbol of the area's economic rejuvenation. Working closely with local government and NMP GmbH & Co. KG (the limited partnership formed by Stadium Developments Ltd. and joint venture partner P&O), RTKL created a plan that reflects both their commercial aspirations and the exacting requirements of the brief. The result is a definitive master-plan that provides a framework for the 83 hectare urban regeneration development. 'Working as part of the CentrO team has been an extremely fulfilling experience for us,' says Paul Hanegraaf, Managing Director of RTKL UK Ltd. 'We have witnessed the transformation of this industrial landscape into the next wave of integrated retail and leisure development'.

At the heart of the project is a 230-store retail centre complemented by a variety of leisure components including a waterfront promenade anchored by an 11,000-seat arena and an RTKL-designed nine-screen Warner cinema. Other features include a Planet Hollywood Restaurant, CentrOPark children's activity park and a 1,200-seat themed 'Oasis' food court. The development expects to attract over 30 million visitors a year and marks the beginning of a new commercial era for the surrounding community. The lighting design for the retail centre was in the hands of Theo Kondos and Associates, New York.

The centre court was created by RTKL as an oval. All the adjacent ceilings are modulated with lit coves, and a catwalk follows the oval skylight which holds the accent and functional lighting for the centre court. A hint of green neon in the lozenge-shaped cupola gives that extra touch of theatre *(facing)*.

The food court, named the 'Oasis', was designed as a street scene with lit windows and a continuous blue perimeter which accentuates and mimics an evening sky. The 'Oasis' is a comfortable meeting space and always has atmosphere and ambience *(left)*.

The typical mall section is well lit, but more important, it is accented with lines of light on the truss, pockets of light in the clerestory windows, accent lighting on the trees, and a circle of light at the transition courts *(left, below)*.

Commodity, Firmness and Delight

Good lighting makes that object of desire more desirable – in a shop window, in a display case, on a shelf. That is the excitement of retail lighting, and why it is one of the busiest professional areas for lighting designers. The opportunity to animate a new environment, to make what is on display appear even more interesting and welcoming to the customer, is a fascinating challenge that requires both sensitivity to the nuances of the space on the part of the designer, and good technical knowledge.

It is an open question whether lighting design is an art or a science. It is easy to understand this dilemma, for light itself, though measurable and quantifiable scientifically, is also for all of us a personal, sensory experience, felt directly through our eyes, affecting our moods. In my view the exercise of any profession requires a mixture of both knowledge and sensibility, awareness and expertise, facts and visions. And while not pleading a special case for lighting, I do believe that achieving good lighting is not only a worthwhile goal for the designer but also contributes to the well-being of others who use or work in the spaces they light.

As to science, the lighting industry is going through exciting times. New lamps and new fittings offer more opportunities, smaller, safer and faster control gear creates new challenges. All these developments need to be studied and learnt. One of the most interesting aspects of all these developments is the change of scale: fittings are smaller, lamps are smaller, transformers are smaller. (In the last decade, it seems, spotlight housings have reduced from buckets to bare wires.) The output and efficiency of lamps is increasing, meaning we can, as Mies van der Rohe told us, do more with less.

As to art, the important thing is training the eye. This may sound pretentious – like preparing to be an antiques connoisseur or a racehorse tipster – but what I mean is no more than the simple enjoyment of looking and remembering: working out for yourself how this problem was solved, how that effect was reached. And asking yourself how would I have done it, does the solution work, what are its drawbacks, what might be the alternatives. Whenever and wherever I travel, one of the pleasures is to look at other designers' work and see how they have done. Talking to architects and designers, to retailers and

managers is another source of information and approaches. Looking and learning in these ways will help you to meet the challenges of each new brief with greater awareness, greater flexibility and more enjoyment. One of the pleasures of designing is working with the client. Raymond Loewy, the Franco-American product and graphics designer, once described the designer's job as 'giving the client what he didn't think he could get'. Sometimes this can happen, and a design can exceed anybody's expectations. Such occasions are rare, but there is real joy in creating a solution that you know is successful and unique. More often it is a matter of listening to the client, looking at the brief, working with the architect and construction team, and creating a solution that does the right task in an effective and attractive way. Most designs involve

compromise, but that should not be seen as a failure, rather as a product of teamwork. Often there is not the time or the budget for an ideal solution, and the designer has to do the best possible with what is available.

The Roman architectural writer Vitruvius describes the business of architecture as 'commodity, firmness and delight'. By commodity he meant the fitness of the solution to its purpose, by firmness the design's enduring strength and stability, and by delight the pleasure given to the maker and the user. The same motto could stand for lighting design. A good design's desirable qualities should be its sense of rightness, its technical validity as an installation, and the comfort and enjoyment brought to those it lights.

Glossary

Adaptation: The process which takes place as the visual system adjusts itself to the brightness or the colour (chromatic adaptation) of the visual field. The term is also used, usually qualified, to denote the final stage of this process.

Apparent colour: Of a light source; subjectively the hue of the source or of a white surface illuminated by the source; the degree of warmth associated with the source colour. Lamps of low correlated colour temperatures are usually described as having a warm apparent colour, and lamps of high correlated colour temperature as having a cold apparent colour.

Average illuminance (Eave): The arithmetic mean illuminance over the specified surface.

Brightness: The subjective response to luminance in the field of view dependent upon the adaptation of the eye.

Candela (cd): The SI unit of luminous intensity, equal to one lumen per steradian.

Chroma: In the Munsell system, an index of saturation of colour ranging from zero for neutral grey to ten or over for strong colours. A low chroma implies a pastel shade.

Colour constancy: The condition resulting from the process of chromatic adaptation whereby the colour of objects is not perceived to change greatly under a wide range of lighting conditions both in terms of colour quality and luminance.

Colour rendering: A general expression for the appearance of surface colours when illuminated by light from a given source compared, consciously or unconsciously, with their appearance under light from some reference source. Good colour rendering implies similarity of appearance to that under an acceptable light source, such as daylight. Typical areas requiring good or excellent colour rendering are quality control areas and laboratories where colour evaluation takes place.

Colour temperature (Tc, unit: K): The temperature of a full radiator which emits radiation of the same chromaticity as the radiator being considered.

Contrast: A term that is used subjectively and objectively. Subjectively it describes the difference in appearance of two parts of a visual field seen simultaneously or successively. The difference may be one of brightness or colour, or both. Objectively, the term expresses the luminance difference between the two parts of the field.

Diffuse reflection: Reflection in which the reflected light is diffused and there is no significant specular reflection, as from a matt paint.

Directional lighting: Lighting designed to illuminate a task or surface predominantly from one direction.

Discharge lamp: A lamp in which the light is produced either directly or by the excitation of phosphors by an electric discharge through a gas, a metal vapour or a mixture of several gases and vapours.

Downlighter: Direct lighting luminaires from which light is emitted only within relatively small angles to the downward vertical.

Fluorescent lamp: This category of lamps functions by converting ultraviolet energy (created by an electrical discharge in mercury vapour) into visible light through interaction with the phosphor coating of the tube.

Glare: The discomfort or impairment of vision experienced when parts of the visual field are excessively bright in relation to the general surroundings.

Hue: Colour in the sense of red, or yellow or green etc. (See also Munsell.)

Illuminance (E, units: lm/m2, lux): The luminous flux density at a surface, i.e. the luminous flux incident per unit area. This quantity was formerly known as the illumination value or illumination level.

Incandescent lamp: A lamp in which light is produced by a filament heated to incandescence by the passage of an electric current.

Lumen (lm): The SI unit of luminous flux, used in describing a quantity of light emitted by a source or received by a surface. A small source which has a uniform luminous intensity of one candela emits a total of 4 x pi lumens in all directions and emits one lumen within a unit solid angle, i.e. 1 steradian.

Luminance (L, unit: cd/m2): The physical measure of the stimulus which produces the sensation of brightness measured by the luminous intensity of the light emitted or reflected in a given direction from a surface element, divided by the projected area of the element in the same direction. The SI unit of luminance is the candela per square metre.

Lux (lux): The SI unit of illuminance, equal to one lumen per square metre (lm/m2).

Munsell system: A system of surface colour classification using uniform colour scales of hue, value and chroma. A typical Munsell designation of a colour is 7.5 BG6/2, where 7.5 BG (blue green) is the hue reference, 6 is the value and 2 is the chroma reference number.

Optical radiation: That part of the electromagnetic spectrum from 100nm to 1mm.

Purity: A measure of the proportions of the amounts of the monochromatic and specified achromatic light stimuli that, when additively mixed, match the colour stimulus. The proportions can be measured in different ways yielding either colorimetric purity or excitation purity.

Reflectance (factor) (R, p): The ratio of the luminous flux reflected from a surface to the luminous flux incident on it. Except for matt surfaces, reflectance depends on how the surface is illuminated but especially on the direction of the incident light and its spectral distribution. The value is always less than unity and is expressed as either a decimal or as a percentage.

Saturation: The subjective estimate of the amount of pure chromatic colour present in a sample, judged in proportion to its brightness.

Uplighter: Luminaires which direct most of the light upwards on to the ceiling or upper walls in order to illuminate the working plane by reflection.

Utilisation factor (UF): The proportion of the luminous flux emitted by the lamps which reaches the working plane.

Value: In the Munsell system, an index of the lightness of a surface ranging from 0 (black) to 10 (white). Approximately related to percentage reflectance by the relationship $R = V(V-1)$ where R is reflectance (%) and V is value.

Working plane: The horizontal, vertical, or inclined plane in which the visual task lies. If no information is available, the working plane may be considered to be horizontal and at 0.8m above the floor.

Index